Introduction to SPC

Donald Holmes
Stochos Inc. and Union College

Copley Publishing Group
Littleton Massachusetts 01460

Printed in the United States of America.

ISBN 0-87411-248-6

Contents

Introduction

You have just purchased an SPC program - I hope it is Custom/QC of course - and you have begun to get some output. The questions now start to pile up about what does all of this stuff mean to you.

The purpose of this text is to explain the various items that come out of the SPC type of statistical analysis programs. At times I will wax philosophical when I come to some points where experience indicates some confusion exists. At other points I will wax philosophical anyway.

First I have to mention several types of data that you are likely to meet as you work to describe and to control the output of your process.

Examples of such data include:

- diameters
- lengths
- weights
- densities
- percent weight lost
- percent of product defective
- average number of impurities per item

The first five examples are the results of measurements taken on the items being manufactured and are referred to as "variables data". The last two examples are the results of counting activities with respect to the items. The data in this case is referred to as "attributes data". The terms "continuous" and "discrete" are also used, respectively, to denote these data types.

Another term used is "nominal". Nominal data is the type which, for example, describes the types of defects in manufactured products: dents, scratches, nicks, holes, neps, etc. There is no implied order in this data - nicks are not greater than neps. This use of the word "nominal" should not be confused with the nominal value of process which is used in the context of specification limits. This is the type of data which is used in Pareto analysis which I will describe a little later.

I will assume, as you will see soon, that you are familiar with such ideas as:

Average Value - it's the IQ score that your children's scores are above and your neighbor's children are below. It is sometimes referred to as the "mean" value.

Normal Curve - it's the way your tests were graded when you thought an 88 was good for a B grade and your teacher gave you a D. It is also called a Bell Curve or a Gaussian Curve. We will use the word "Normal". Diameters, lengths, and weights are usually Normally distributed.

Probability - a way to describe the chances that your picnic will be rained on. I suspect that you already know that the probability of drawing an ace from a deck of well mixed cards is 4 in 52 - or 4/52. It is probably further back in your memory that the probability of drawing an ace and a king is 2(4/52)(4/51) - but, the probability that this will keep you from understanding the rest of the text is about .0001.

Binomial Curve - that's the one that tells you the probability that 4 out of the next 5 of your picnics will get rained on - given your kind of luck. The number of bad items found in a sample of 10 items is usually assumed to be Binomially distributed. (Note my use of the word "bad" in place of the awkward "nonconforming" that is being advocated these days. My personal belief is that the length of the words we use to describe our predicament is directly proportional to the size of the problem we face. The increased word length is used as a security blanket. So, I'll just be nonconforming and use the terms "good" and "bad".)

When the assumptions above become important, I will give some references that you may use to refresh your memory.

Now let's get on with the task of clarifying the terms used in computer outputs.

Descriptive Statistics (for ordinal data):

Usually the first segment of the output from an SPC package is a description of the output of a process. The question as to whether or not that process was stable during the data collection period is addressed in the process control segment of the package.

This section will cover:

- Histograms
- Mean, Median, Mode
- Range, Standard Deviation
- Skewness, Kurtosis
- Process Stability
- Process Capability Measures - CP, etc.

The "description" of the output of the process is conceived of as a curve which represents how often each of the various data values will occur as the process continues. This curve is referred to as the "frequency curve" for the process. The shape of the frequency curve is dependent on the process being studied. For variable data gathered from a mechanical process, such as part diameters, the shape of the curve is approximately "Normal" ("Bell shaped" or "Gaussian"). For variable data gathered from a chemical process, such as concentration of chemical components, the curve is often not Normal.

The shapes of the curves for attribute data depend on the question being asked about the process. The "Binomial" is one which is frequently used to describe the number of bad items which occur in a random sample from a process. The "Geometric" is one which describes how many subgroups will be drawn from a process before you get an out-of-control signal.

Histograms and Related Ideas:

The description of the output of the process generally begins with a graphical display which indicates the number of data points falling within given limits. An example for the densities of pressed powder parts is shown below in Figure 1.

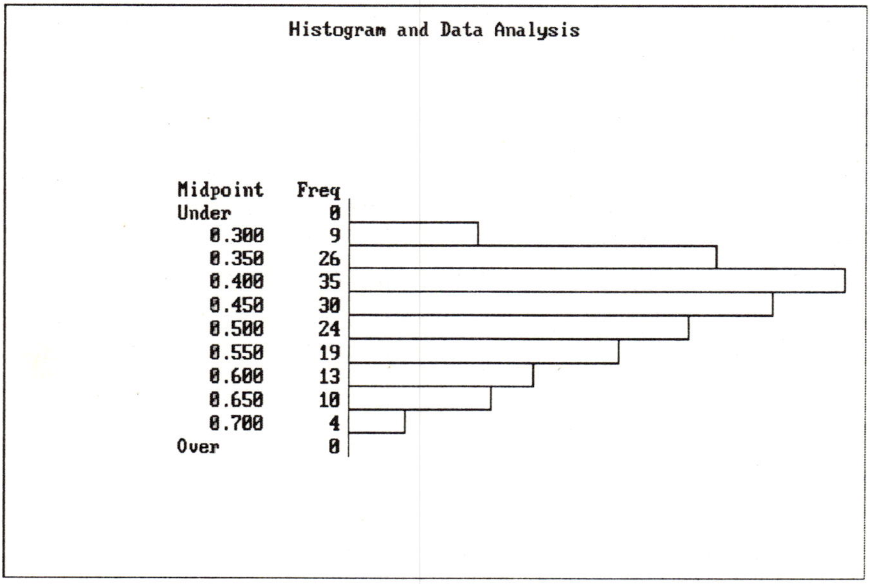

Figure 1

This graphical display is referred to as a "Frequency Histogram", or just as a "Histogram". There are 11 "Classes" in this histogram. There are 2 "open" classes - the first and the eleventh. There are 9 "closed" classes - those whose "Class Midpoints" are: 0.300, 0.350, 0.400, 0.450, 0.500, 0.550, 0.600, 0.650 and 0.700. The class boundaries for the second class (the one whose midpoint is 0.300) are 0.275 and 0.325. The lower class boundary for the second class is the upper class boundary for the first class - which has no lower class boundary. That's why the first class is referred to as an open class. The 11th class is also an open class - it has a lower but not an upper class boundary. The lengths of the rectangles erected on the class boundaries are proportional to the number of data points that fell in between the class boundaries. Thus there were 35 items whose values were between 0.375 and 0.425.

The data from 170 parts used in constructing this histogram are shown below in the "Frequency Table".

Frequency Table
for
Part Densities

Boundaries	Midpoints	Frequency	Relative Frequency	Cumulative Relative Frequency
.275 - .325	.30	9	9/170 = 0.05	0.05
.325 - .375	.35	26	26/170 = 0.15	0.20
.375 - .425	.40	35	35/170 = 0.21	0.41
.425 - .475	.45	30	30/170 = 0.18	0.59
.475 - .525	.50	24	24/170 = 0.14	0.73
.525 - .575	.55	19	19/170 = 0.11	0.84
.575 - .625	.60	13	13/170 = 0.08	0.92
.625 - .675	.65	10	10/170 = 0.06	0.98
.675 - .725	.70	4	4/170 = 0.02	1.00
		170	170/170 = 1.00	

The "Frequency" column indicates the number of parts whose densities fell within the specified class boundaries. A frequency in this column is sometimes referred to as the "Absolute Frequency" so as to distinguish it from those in the "Relative Frequency" column, which represent the class frequencies divided by the total frequency. Multiplying the relative frequency by 100 yields the percentage of parts within each class.

The "Cumulative Relative Frequency" times 100 indicates the percentage of the 170 parts whose densities were less than or equal to the "Upper Class Boundary".

Knowledge of the process frequency curve enables you to establish the percentage of your product which meets specifications - or to determine the effects of changing the spec limits on the percentage of parts which would meet the new specs.

WARNING !

The histogram of the data is a way to obtain an estimate of the frequency curve. Obtaining reasonably good estimates of the curve shape, however, requires rather large sample sizes - about 3000 - to bracket the cumulative curve within an absolute error of about +/- 5%. (See Dixon and Massey(1), page 345, for details on how to set up this interval.)

BASED ON THIS FACT YOU SHOULD BE RELUCTANT TO DRAW CONCLUSIONS ABOUT THE SHAPE OF THE CURVE IF YOU HAVE MADE A HISTOGRAM FROM 50 DATA POINTS!

END OF WARNING !

Numerical Description of the Properties of a Frequency Curve:

Another way to approach the question of the shape of the frequency curve is to use calculations on the data which help to describe the curve.

PROCESS CENTER:

The first descriptive property of the curve is its center. That is, is the curve of the part densities centered around 0.50 or 0.60?

There are a number of measures of the center of a process, but I will consider only the three most common: the mean (also referred to in this text as the "average"), the median, and the mode.

Mean:

The Mean value is the sum of all of the measurements divided by the number of measurements that went into the total. In SPC terminology, the individual measurements are denoted by "X", and the average value is denoted as "Xbar". Thus, the average value is calculated as:

Xbar = Sum of X's / Number of X's

For example, suppose the X's are 5,4,6,7,3. Then the average is 25/5 = 5. Now I know that you know that, but I just couldn't omit it.

Median:

The Median value is the value which has as many data points above as below it once the data has been arranged in ascending order of magnitude. For example, suppose the X's are (in order of collection) 5,4,6,7,3. Arranging the data in order, we have 3,4,5,6,7. The Median for this data is then 5. In descriptive statistics, the median is often the measure used to describe the center of a frequency curve which is not symmetric. In the example the median and the mean have the same value, but this is not generally true.

Mode:

The Mode is the most frequently occurring value. For example, suppose the X's are 2,3,4,4,3,5,4,6,3,7,5,8 6,4. The value 4 occurs more often than any other in the data and hence the mode is 4. In complex curves, there may be several distinct peaks in the curve - each of which is referred to as a mode.

PROCESS WIDTH:

The next descriptive property of a curve is its width. This property is used to answer such questions as "Do the densities of the products from press number 1 vary as much as the products from press number 2?" or, "Is the histogram for part A wider than the histogram for part B?"

There are also many additional measures of width in use. I will discuss three of these measures: Range, Standard Deviation, and Square Root of the Mean Square of the Successive Differences.

Range:

The Range is defined as the distance between the largest X and the smallest X. Thus, in the data 5,6,3,4,7,5, the range is 7 - 3 = 4. This is certainly the simplest measure of width but also the least effective since all but two of the data points are ignored. The data 3,3,3,4,5,6,7,7,7 for example has the same range as the previous data. Yet, you have an intuitive feeling that the second data set is wider than the first data set. The following measure takes all of the points into consideration in its calculation.

Standard Deviation:

The Standard Deviation is the square root of the average of the square of the distances that the X's are from the process average. For example, suppose the data are 5,6,4,7,3. The average of this data is 5. The remainder of the calculation of the standard deviation is shown below.

Step 1: Calculate the Average (Xbar)
Xbar = (5+6+4+7+3)/5 = 5

	X	Step 2: Subtract Xbar from each X X-Average	Step 3: Square the differences (X-Average)^2
	5	5-5 = 0	0* 0 = 0
	6	6-5 = 1	1* 1 = 1
	4	4-5 = -1	-1*-1 = 1
	7	7-5 = 2	2* 2 = 4
	3	3-5 = -2	-2*-2 = 4
Step 4: Get totals		0	10

Step 5: Get average of differences squared - 10/5 = 2

Step 6: Take square root of the average of the squared distances

Sqr(2) = 1.414

And thus, the standard deviation is 1.414.

Now this measure is more complex to calculate but has much more power than the Range. And, the complexity of the calculation is no longer an issue now that you have an SPC package for your microcomputer.

In summary, the standard deviation is an average distance that data items are from the process center. Admittedly, it is a funny kind of average, but that's all it is. You engineers have heard it referred to as an "RMS value". You should note that the sum of the deviations from the average value is zero. This is a fundamental property of average values.

I mustn't forget to mention "Divide by n or divide by n-1?". I divided by n so as not to obscure the fact that the standard deviation is really just an average value. The use of n-1 rather than n is an attempt to improve the estimate of the standard deviation when the sample size is small. In SPC work, the value obtained using n is referred to as "sigma"; the value obtained using n-1 is referred to as "s". In the "Histogram and Data Analyis" segment of Custom/QC, we use n-1 as the denominator. In the process control segment, we use n as the denominator. This seems to be more or less the standard industry practice. Neither is particularly right or wrong - it just helps to know which one is being used.

Standardization of nomenclature is coming! A ready reference in this area is ASTM's standard on data analysis and presentation - STP15. I have stayed pretty much with the notation of the classic texts - Duncan's and Grant's.

IMPORTANT NOTE !

THE AVERAGE AND THE STANDARD DEVIATION ARE THE BASIC INGREDIENTS OF ALMOST ALL OF THE ANALYSIS WORK DONE IN THE AREA OF STATISTICAL PROCESS CONTROL!

END OF IMPORTANT NOTE !

Mean Square Successive Difference:

The next measure, the Square Root of the Mean Square of the Successive Differences, is included because it is useful in the concept of Process Capability. (Please see the section on PROCESS CAPABILITY.) Although the calculation may seem to be different than that used in the standard deviation calculation above, the end result is in fact a standard deviation.

This measure has a long title, so its calculation is demonstrated to firm up the definition. Suppose the data set is 7,5,6,3,4,2,4,5.

	Step 1: Take the Successive Differences	Step 2: Square the Successive Differences
X	(X - Last X)	(X - Last X)^2
7	N.A.	
5	-2	4
6	+1	1
3	-3	9
4	+1	1
2	-2	4
4	+2	4
5	+1	1
Step 3: Get the total		24

Step 4: Take the average - 24/7 = 3.429

Thus the Mean Squared Successive Difference is 24/7 = 3.429.

Step 5: Take the square root of the Mean Squared Successive Difference - 1.310.

This measure, when divided by 1.414, is another estimate of the standard deviation of the process - 1.310/1.414 = .927.

"Wait a minute", you say, "Which one is the 'true' standard deviation?". Well, let's see if I can clarify the situation with an example. Suppose the data is as follows: 1,2,3,4,5,6,7. The width of the process, as measured by the usual standard deviation, is SQR(28/7) = 2. This measure of the process width has been greatly influenced by the fact that there is a definite upward trend in the data. For this data the successive differences are all 1's and hence the value of the square root of the mean square successive differences estimator of the standard deviation is 1/1.414 = .707. The first standard deviation is large because of the trend. The second one is small because it ignores the trend. Thus, the second one is an indication of what the standard deviation of the process would be if there were no changes in average value. We include it as a part of the Custom/QC descriptive statistics segment so that you have an indication as to whether or not the process is stable. If the process indicates lack of stability, then you should be cautious about using the reported mean and standard deviation as measures of process performance.

You really have to get used to this idea of different standard deviations. It is basic to understanding what goes on in SPC analysis.

The Mean Square Successive Difference (MSSD) concept is very similar in nature to the "moving range" which I will talk about when discussing X-Moving Range control charts.

So, the first standard deviation is a measure of what the process is doing - Process Doability. The second is what the process is capable of doing - Process Capability. The second standard deviation is an estimate of the best your process will be able to do with respect to its variability.

An item that you see reported in Custom/QC is the "Normal Z" for the comparison of the two standard deviations. If this number is greater than about 3, then you have the advance indication that the process is not stable. If you use the "Histogram and Data Analysis" segment of Custom/QC you will see that we report several process capability indices (see PROCESS CAPABILITY below) - one of which uses the MSSD standard deviation described above. That way you have an early estimate of the potential capability of your process. (See Dixon and Massey(1) pages 406-407 for more information on the calculation and interpretation of both the standard deviation and the Z value.)

SYMMETRY - or lack of same - SKEWNESS:

The next property of a frequency curve deals with the question of the symmetry of the curve relative to its center - as measured by the average. A "Normal" curve is symmetric about the mean value. The measure of symmetry, called the "Skewness", is zero for such a curve. The skewness for a curve such as that shown in Figure 1 is positive. Please note my informality - here's Figure 1 again where you need it. Should I have called it 2 or let you turn back several pages?

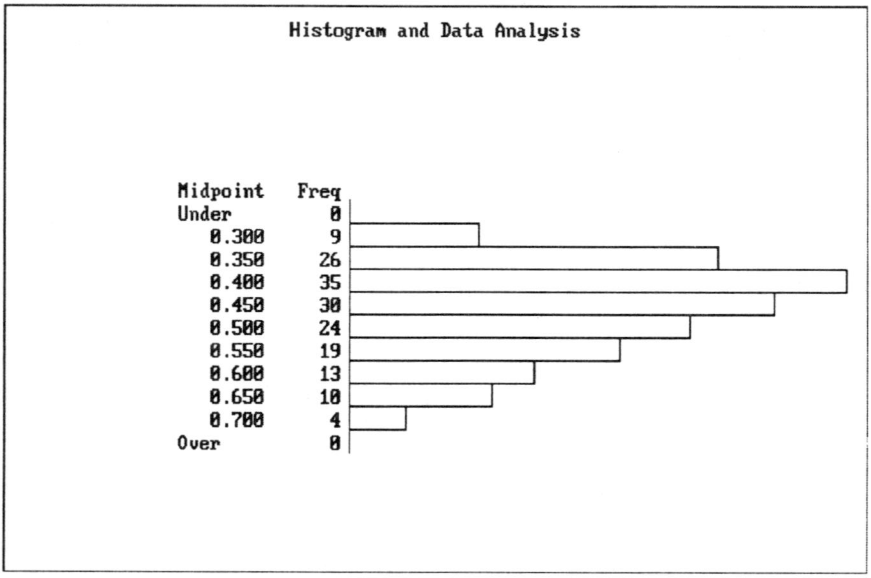

Figure 1

A curve with the tail off toward the small X's has negative skewness. Again, there are several measures of skewness in use. One is based on the fact that the mean and the median, for a curve which is symmetric, will be equal. Hence the one measure is (Mean - Mode)/Sigma. Figure 2 below is a curve with negative skewness.

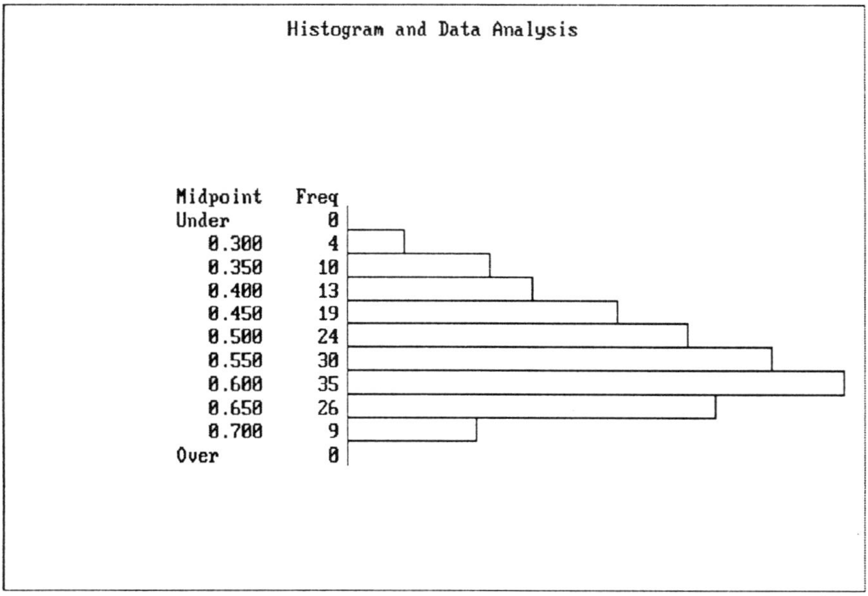

Figure 2

A more commonly used measure is the average of the cubed deviations of the X's from the mean divided by the standard deviation cubed. This measure is sometimes referred to as the "third moment measure" of skewness. The third moment method is the one used in Custom/QC.

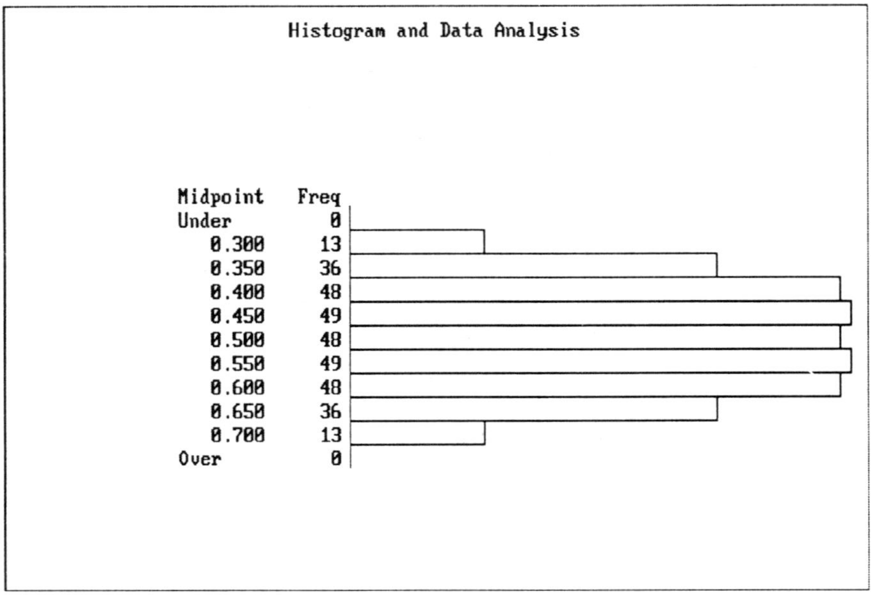

Figure 3

The skewness reported in the descriptive statistics segment should be close to zero for your curve to be considered as "Normal". The data displayed in Figure 3 above has a skewness of zero - it is symmetric - but it is not very Normal.

KURTOSIS:

This measure describes the tendency of a curve to have long tails and a high center. For a Normal curve, this measure is 3. (There is another measure sometimes referred to as "excess" which is simply the kurtosis minus 3.) For a curve that is more pointed or has longer tails than a Normal curve, the kurtosis is greater than 3. For flatter curves, the kurtosis is less than 3. Thus for Figure 3 above, which does not have the long tails of a Normal, we expect the kurtosis to be less than 3. (It is actually about 2.)

A histogram for a curve more pointed than a Normal is shown in Figure 4. The value of kurtosis for this curve is about 4.

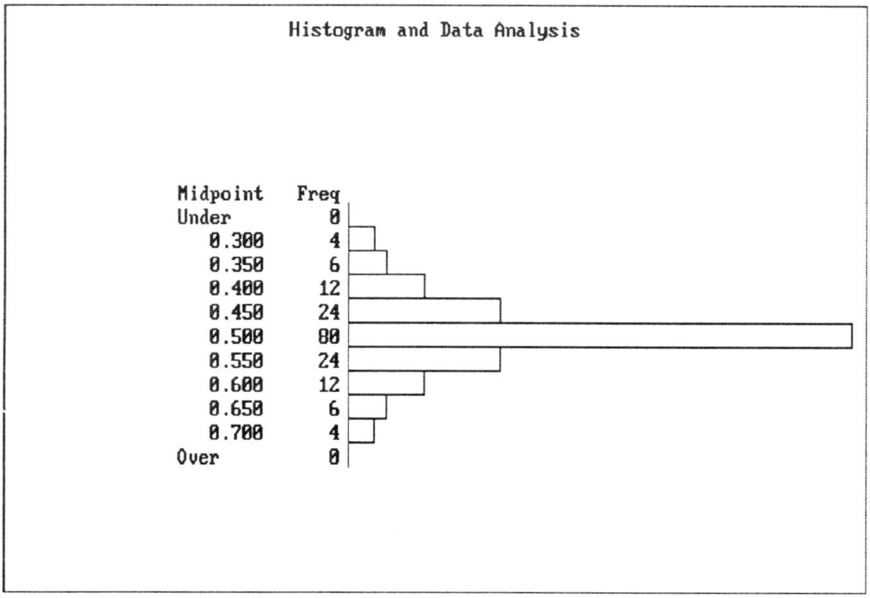

Figure 4

The measure used for kurtosis is the mean of the fourth power of the deviations from the mean divided by the fourth power of the standard deviation.

The measures of Skewness and Kurtosis are primarily used to make a quick assessment of the Normality of your curve. For a rough assessment of Normality, the skewness should be about 0 +/- 7.35/Sqr(n) and the kurtosis should be about 3 +/- 14.70/Sqr(n). Measures of skewness or kurtosis outside these values is an indication that the frequency curve is not Normal.

EARLY WARNING OF PROCESS INSTABILITY:

We have elected in Custom/QC to give you several indications of the dynamic performance of the process in the descriptive statistics segment of the program. I have discussed the successive difference standard deviation estimate above and its implications for determining the stability of your process.

ANOTHER TEST FOR PROCESS STABILITY:

The Run Test checks to see how many times the data lies in lumps of data which are either all below the median or all above the median. Such a lump is more elegantly called a "run".

Let's take several examples to clarify the concept. First, consider the data we used above: 1,2,3,4,5,6,7. The median is 4. The first three values are all below 4 and constitute a run below the median. The last three values are all above the median and constitute a run above the median. Thus the data has two runs.

As an example of another type of nonrandomness, consider the data: 3,5,3,5,3,5,3,5. The median is again 4.

The number of runs is 8. Thus you can see that for cyclic data the number of runs is large and for trend data, the number of runs is small. Custom/QC reports 95% and 99% limits for the number of runs your data can demonstrate without your having to be concerned about the validity of the descriptive statistics measures reported. An observed number of runs outside of these limits is an indication that the underlying process is unstable and that the descriptive measures should be used with caution for estimating future process performance.

Process Capability:

Once the process does NOT give indications of being UNstable, the average and standard deviation can be used to provide estimates of the center and width of the process. The results can then be compared with the desired process center (the nominal value), and the desired process width (the tolerance).

ESTIMATION OF PERCENTAGE OF PRODUCT INSIDE SPECIFICATIONS:

I assume that you are familiar with the idea of specifications. I will refer to the upper spec limit for the process as USL and the lower spec limit as LSL. I will use the word "tolerance" to mean USL - LSL.

If the curve of the process is Normal, then "Normal Tables" such as those found in statistics texts or in the "Frequency Curve Comparison" segment of Custom/QC, can be used to determine the percentage of products lying inside any given limits.

To do this, the average and the standard deviation are used to form a standardized variable known as Z. The value of Z for any specified X is determined by subtracting the average from X and then dividing the result by the standard deviation. For example, suppose the average of the process is 50 and the standard deviation is 2, then the Z value for X = 54 is (54-50)/2 = 2. Values of Z are positive if the X value is above the average and negative if the X value is below the average.

The Z value is a number which indicates the distance an X is from the average value - in terms of multiples of the standard deviation.

The term "outside three sigma" refers to data points, in terms of the example, that are above 56 or below 44 - average plus and minus 3*sigma. The value of Z in this case is 3.

The tables for the Normal curve indicate that 99.7% of all data will lie inside the 3 sigma limits. Another often used fact is that 95% of all data will lie inside the 2 sigma limits. Since 99.7% of all data lies within 3 sigma limits, the actual width of your process is taken to be about 6 sigma - 3 sigma above the average and 3 sigma below the average.

If your process is not Normal, all is not lost! Call your friendly statistician for help. Ask him about Chebychev's Inequality, log transformations and stuff like that.

PROCESS CAPABILITY MEASURES:

This section will deal with the Taguchi measures of "Process Capability". But, before I do that, please let's review the ideas of specifications.

NOMINAL - the value for which the product has the highest quality.

LOWER SPEC LIMIT - the smallest value for which the product is just acceptable. You don't want many of these; they are poor quality.

UPPER SPEC LIMIT - the largest value for which the product is just acceptable. You don't want many of these; they are poor quality.

Those who disagree with the concept that products near the spec limits are of poorer quality than those near the nominal have probably led a sheltered life - and never had to use the products they produced. Just ask those who have!

Taguchi formalizes this idea by saying that the dollar loss due to a product not being made to nominal is proportional to the square of the distance the product is from the nominal. With this point of view in mind, he uses a measure - which he unfortunately called the "standard deviation" - which is based on deviations from nominal rather than deviations from average. This means that his standard deviation is larger than the ordinary one if the process is not centered on the nominal value. The following relation holds:

Taguchi's Sigma Squared = Usual Sigma Squared +
(Nominal - Average) Squared

So, you can see that his standard deviation includes both the width of the process and the distance the process is from nominal.

(Stochos has built a Quality Portfolio chart based on this relationship. Call us for copies if you are interested, or see Quality - Dec 1986, page 67.)

All of the measures discussed in the following sections are in terms of sigma. You must decide for yourself which sigma you wish to use. (Custom/QC reports the measures using both definitions of sigma.)

COMPARISON OF DESIRED PROCESS WIDTH WITH OBSERVED PROCESS WIDTH:

The first measure of process capability is the "CP". This measure is defined as the desired width (tolerance) divided by the actual width (6*sigma). The goal for this measure is for it to be greater than 1.33. Another way to express this goal is that the tolerance should be 8*sigma in order for the bad parts produced to be in the "Parts per Million" category: +/- 4*sigma yields 99.994% inside the specs for a Normal curve. Once more:

$$CP = \frac{USL - LSL}{6 * Sigma}$$

Another version of this index is the "CPR" which is just the reciprocal of the CP index. The goal for this index is that it be 0.75 or smaller. The term "CR" is also used for the CPR.

Another quite direct measure of the capability of the process is an estimate of the percentage of products falling within specs. This measure is based on a Normal curve.

COMPARISON OF DESIRED PROCESS CENTER WITH OBSERVED PROCESS CENTER:

(Nominal) (Average)

The next matter for consideration is how close the process average is to the nominal. Assuming that the nominal value is halfway between the spec limits, the position of the average relative to the nominal can be addressed by seeing how close the average is to each spec limit. Six measures are in common use to establish the position of the average relative to the nominal:

Z-upper = (USL-average)/sigma
Z-lower = (LSL-average)/sigma
Z-min = the minimum of the absolute values of Z-upper and Z-lower.
CPU = Z-upper divided by 3
CPL = The negative of Z-lower divided by 3
CPK = The minimum of CPU and CPL

(Note that CP = (Z-upper - Z-lower)/6)

In order to make the interpretation of these values more clear, consider the case diagrammed below. Assume the following:

Xbar = 50, Sigma = 2, USL =57, LSL = 47, Nominal = 52

The plot below shows how the Z values compare with the X values.

```
      LSL              Ave   Nominal               USL
-----!-------------!-------!-----------------!------X scale
      47              50      52                57

-----!-------------!-------!-----------------!------Z scale
    -1.5              0       1                3.5
   Z-lower                  Z-Nominal        Z-upper
          (Taguchi's Zmin would be = 1.5)

-----!---------------------------------------!------% scale
    6.68                                     99.98
          (Hence about 93.3% inside specs)
```

Next I will stretch your brain by putting the CP values on the same plot - but using a "CP" scale rather than X or Z.

```
-----!---------------------------------------!------CP scale
      .5                                      1.16
         CPL = Z-lower/3        CPU = Z-upper/3
         CPK = Z-min/3
```

Z-min being 1.5 (numerically - not algebraically) means that the LSL is closer to the process center than is the USL whose Z is 3.5. Thus, assuming that the nominal value is halfway between the USL and the LSL, the center of the process is not on the nominal value. Z-upper being +3.5 indicates that the USL is 3.5 sigmas above the process center - not far enough away to meet the goal of 4, mentioned in the CP discussion. The situation relative to the LSL is much worse - the distance is only 1.5 sigmas. The process is poorly centered relative to the nominal value. Z-upper would have to be numerically equal to Z-lower for the process to be centered on the nominal value.

The description of where the process is centered based on the CPU, CPL, Z-upper, Z-lower and nominal being halfway between specs seems very strained to me. So, I slipped in one other Z value: Z-nominal. With Z-upper, Z-nominal, and Z-lower the situation is very clear - whether or not the nominal is halfway between the specs and whether or not the specs are one sided. Z-nominal, for the example, is equal to 1. This clearly shows that the nominal is one standard deviation above the process average.

The values of Z-upper, Z-lower, and Z-nominal thus contain all of the information needed to determine the answers to both questions of width relative to tolerance and center relative to nominal - and you can forget the CP stuff. Now I realize that you must use CP stuff because your customers require them. But, perhaps we can work together for rationality.

COMPARISON OF SEVERAL PROCESSES
(or Same Process at Different Times)

Since the values of the Z's are dimensionless, they can be used to compare the status of different processes or the same process at different points in time. The "Process Capability" segment of Custom/QC contains a CP chart which plots Z-upper, Z-lower, and Z-nominal.

Since all of the process capability measures mentioned above are expressable in terms of the three Z's, the CP chart summarizes visually the information contained in the whole gamut of indices mentioned in this section.

DESCRIPTIVE STATISTICS (for nominal data):

Pareto Charts:

A Pareto chart is similar in form to a histogram, but the "x-axis" is nominal rather than ordinal. The "y-axis" of the Pareto chart is the same as the one in the histogram: frequency of occurrence. Since the data is nominal the names of the types of defects, on the x-axis, are ordered using the frequency of occurrence - the most frequent being plotted first. This chart is useful for identifying which types of defects occur most often. Once identified, the overall quality level is improved by concentrating on eliminating the most frequently occurring defects.

This chart is named for an Italian economist. The curve shape that occurs often in practice is such that about 80 percent of the problems are caused by 20 percent of the defect types. This is referred to as the the "80 - 20" rule. Pareto, for example, found that 80 percent of the wealth was controlled by 20 percent of the population - at least in Italy, at that particular time in history.

The Pareto chart in Figure 5 is for data collected from daily records of an industrial clinic - not production defect data. The causes of trips to the clinic are 3 letter codes whose definitions are probably clear. (e.g. LAC = lacerations)

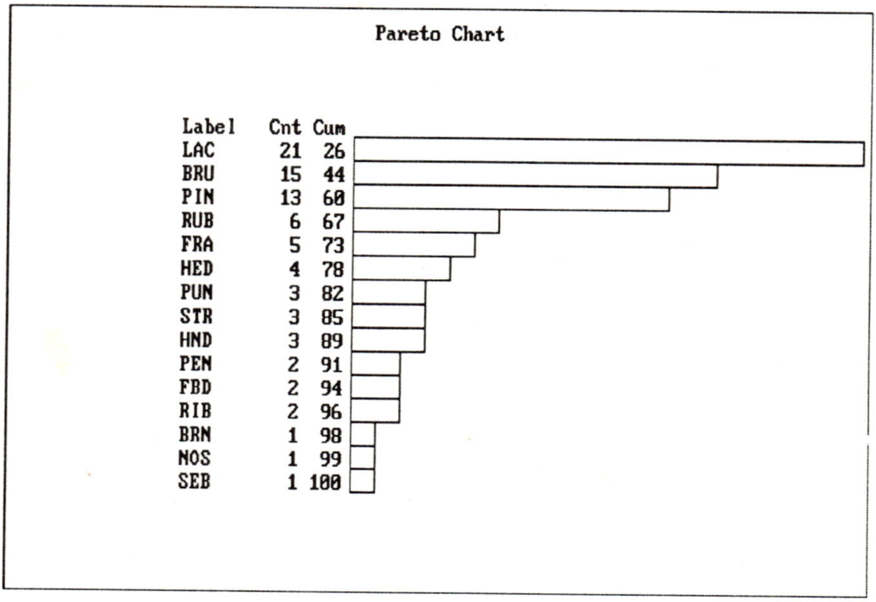

Figure 5

This finishes the discussion of descriptive statistics and we move on to the Process Control aspects of SPC.

Control Charts:

ELEMENTARY CONTROL CHART DEFINITION:

Statistical Process Control is centered around the use of Control Charts. A Control Chart is a graphical representation of (usually) small samples taken to monitor the performance of process variables. The control chart has "Control Limits" which represent the expected range of variation for sample values. If a sample value falls outside the limits, the process is called "out-of-control". Corrective action is taken and the process is brought into control. Once a process is in control, the process capability measures discussed above are used to describe the conditions of the process relative to the specifications.

POTENTIAL RISKS IN USE OF CONTROL CHARTS:

There are risks associated with any sampling procedure - including control charts. There will be times when the process changes and there is no out-of-control signal. There will also be times when the process is in control but the chart says that it is not in control. If the process is out of control and the chart does not detect it, then you are producing a product whose quality is potentially lower than possible until the change is detected. If the process has not changed but the chart says it has, you will either take action which will induce extra process variation or you will perform an analysis only to find that nothing is amiss. Walter Shewhart(2), in his pioneering text, says that his studies indicated that the three sigma limits seemed to strike an economical balance between the two types of risks - for the processes he studied.

RISK TYPE 1 - Looking for trouble when there isn't any:

The usual approach to risk evaluation in process control is to state the risk associated with looking for trouble when there isn't any. That is, 3 sigma limits are used because the chance of hunting for nonexistent problems is limited to about 3 in 1000. With this in mind, other unusual patterns of data have been developed which have approximately the same 3 in 1000 chance of occurring if your process is in control. The rules include:

- 1 point outside of +/- 3 sigma
- 9 points in a row on one side of the average
- 6 points in a row steadily increasing or decreasing - (trend)
- 14 points in a row alternating up and down (cycles)

2 out of 3 points outside of +/- 2 sigma
4 out of 5 points outside of +/- 1 sigma
15 points in a row inside +/- 1 sigma
(probably poor subgroup decision)
8 points in a row NOT inside +/- 1 sigma
7 points in a row on one side of the average
(the probability is not even close to .003
for this one, but the rule is in general use.)

You have the capability when using Custom/QC to use one or more of these rules when evaluating the state of control for your process. YOU SHOULD BE AWARE THAT THE MORE RULES THAT ARE USED IN THE EVALUATION, THE GREATER IS THE PROBABILITY THAT YOU WILL BE HUNTING NON-EXISTENT PROBLEMS. The 3 in 1000 risk increases approximately by a factor of the number of tests used. Thus three tests changes the probability from 3 in 1000 to 9 in 1000 that you will be searching in vain - or worse still that you have changed the process needlessly and introduced excess variability. To see this in what may be a more meaningful light, please read the discussion of Average Run Length below.

The risk evaluation procedure above is rather limited - it tells you nothing about the other side of the risk problem: the process has changed but there is no forthcoming out-of-control signal.

RISK TYPE 2 - Having trouble but not looking for it:

AVERAGE RUN LENGTH AS A RISK ASSESSMENT TOOL:

Leo Aroian(3) developed the concept of using what is now known as the "Average Run Length" (ARL) criterion for describing the risks of sampling that are inherent in the use of control charts. The ARL is the average number of subgroups that you will take before you get the out-of-control message. If you are using 3 sigma limits and the process has not changed, then the average number of subgroups until the signal is 1000/3 = 333: the reciprocal of the probability of getting the signal. If, on the other hand, you are using the 3 sigma limits and the process average moves up to the upper control limit, the probability of a point out of control is now 1/2, and the ARL is 2 - an average of two subgroups will be taken until the out-of-control signal is given by the chart. This is in agreement with what you would expect - the greater the process shift, the sooner the control chart will send the signal. The nice thing about this measure is that it lets you see the impact of subgroup size and subgroup frequency on your costs - not on your probabilities.

Let's revisit the three rules versus one for out-of-control signals and see what the impact is on the Average Run Length (ARL). If the process average has not moved, the ARL for 3 rules is about 111 rather than 333 - you will be hunting for nonexistent trouble 3 times as often. On the other hand, you will be finding existent trouble much more quickly - depending on the rules selected and the severity of the process change.

I mention this measure because you may have heard the same "practical" rule that I have: "Wait for the second point in a row outside 3 sigma". This certainly does stretch out the time between hunting for nonexistent troubles - the ARL would be in the vicinity of 100,000 - but it raises havoc with finding existent troubles. And, it certainly doesn't help the pursuit of the Parts per Million goals that we hear so much about.

TWO USES OF CONTROL CHARTS:

Control Charts were initially designed by Shewhart(2) for process control purposes in about 1926. Then in 1944, General Leslie Simon(4) began to apply the control chart presentation of data to acceptance inspection. His approach was to take small samples, in arbitrary order, from each submitted lot of product and construct a control chart using the data. A "grand lot" consisted of all of the lots for which the data was not out of control. The acceptance or rejection decision was held until the grand lot was formed. In 1957 Richard Freund wrote an article called "Acceptance Control Charts" in which he dealt with the use of charts for acceptance sampling purposes. In order to distinguish between these two different uses of control charts I will refer to PC charts for process control and QA charts for acceptance sampling. (See ACCEPTANCE SAMPLING section in this book.) The sections that follow directly below concentrate on PC charts. I mention this distinction because today, in the flurry of SPC activities, there is a tendency to miss the distinction. That is, QA type charts are being sent to your customers, at their request, and being thought of as PC charts. The QA charts are fundamentally too late for process control. They are acceptance sampling rather than process control.

BACK TO PROCESS CONTROL CHARTS:

Control charts for PC purposes should be designed and run to help uncover unusual events in the manufacturing process so that the causes of the unusual events can be determined and capitalized on to improve the quality of the product you are shipping. A control chart is a "management by exception" device that allows you to concentrate on processes which are not performing as you would expect.

To perform this task, a control chart functions by comparing the variability between subgroups with the variability you would expect given the variability within the subgroups.

XBAR CHARTS:

To clarify the points above, let's get on with a specific case - the Xbar chart. Xbar charts are used for variables data, such as part diameters, to determine whether or not the center of the process is stable.

"Why", you might ask, "don't we just use X charts for this purpose?". There are two very good reasons for using Xbar rather than X:

1. Xbar charts have smaller Average Run Lengths than X charts.

2. Xbar's are approximately Normally Distributed regardless of the shape of the X distribution. This means that the risks associated with the 3 sigma control are known. If we use 3 sigma limits on X charts for non Normal data such as that found in some processes, the risks are not clear. The main bad side effect of this approach is that the process would be over-controlled since you would be getting more out-of-control messages than you should. Another bad side effect of using X charts for nonsymmetric curves is that the extra number of wild goose chases causes the SPC approach to lose credibility in the eyes of the person charged with controlling the process.

The Xbar chart is formed by taking subgroups of some specified size on a routine basis. A good rule for starting a chart is to have about 100 data points - 25 subgroups of 4 or 20 subgroups of 5.

Step 1: Calculate the average for each subgroup - Xbar

Step 2: Calculate the standard deviation for each subgroup - Sigma

Step 3: Calculate the average of the subgroup averages - Xbarbar

Step 4: Calculate the average of the subgroup standard deviations - Sigmabar

At this point, Xbarbar is our best estimate of the process average and sigmabar is our best estimate of the process standard deviation. Well, that's almost right. The value of sigmabar should be divided by the factor c2 - found in the table in Appendix A of this book - in order to be the best estimate of sigma.

From these two values we calculate the limits within which the individual subgroup averages will lie if the process average is stable.

The centerline for the Xbar chart is Xbarbar.

The Upper Control Limit for the Xbar chart (for 3 sigma risk) is set at:

Xbarbar + 3*(sigmabar/c2)*(1/sqr(n))

where c2 is a factor which adjusts for sampling considerations, n is the subgroup size, and sqr(n) is the square root of n.

The Lower Control Limit for the Xbar chart (for 3 sigma risk) is set at:

Xbarbar - 3*(sigmabar/c2)*(1/sqr(n))

Now the reason that the 3 sigma limits are divided by the square root of n is that the limits are for sample averages - and sample averages don't vary as much as the individual values do. Statistical theory shows that the reduction in variability - which we all intuitively expect - is by a factor of 1/sqr(n).

NOTE:

Some folks - not you or I - are inclined to put spec limits on the Xbar chart. This is a bad practice. It implies that there is a connection between the sampling variability of averages and the spec limits. These are two completely unrelated ideas. As I have often told these folks, the only connection between what the process is doing - as indicated by the Xbar chart - and what you would like the process to be doing - as indicated by the specs - is HOPE.

END OF NOTE:

The formulas I have written above for the control limits for the Xbar chart are simplified and tabled in SPC texts to be Xbarbar + A1 * sigmabar and Xbarbar - A1 * sigmabar. In Custom/QC we use the c2 form shown above because it simplifies our task of allowing you to choose the number of sigma's you wish to use for control purposes - some people use 2, some use 3.09, etc., etc. If you prefer to do a manual calculation, Appendix A contains the values of A1 for various subgroup sizes.

Here is an Xbar chart for some data collected from a manufacturing process.

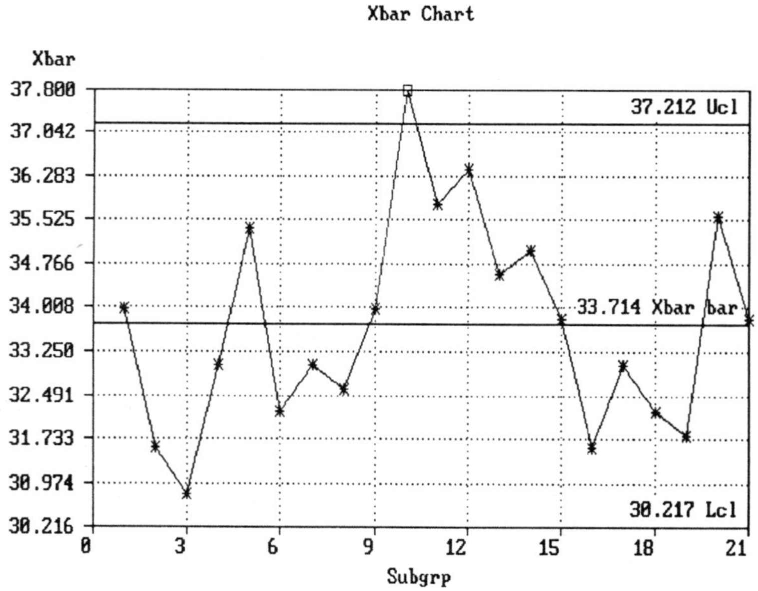

Figure 6

This chart is based on subgroups of size 5. That's a very commonly used subgroup size. It is a throwback to the days of yore when all of the calculations for the control charts were performed by hand. I will talk more about subgroup size later.

The chart above shows out-of-control at subgroup number 10. This is an indication that the process average shifted at that time. An investigation as to the possible cause and/or a control action should be undertaken.

This discussion of Xbar charts demonstrates the general PC chart approach:

- Select samples (subgroups) from an ongoing process.
- Set up expected limits of variation - assuming no process change.
- Plot subgroup measurement on a chart which shows the "control limits." (The table in Appendix A assumes "3 sigma" limits are to be used.)
- If the points stay inside the limits - act as if no process change has occurred.
- If one or more of the points are outside the limits - take action to capitalize on the information. I use the word "capitalize" since a point indicating that the process has significantly shifted in the right direction should not be corrected or considered as bad. Unusual behavior relative to the other decision rules are treated in the same fashion as the control limit rule.

I have used the Xbar chart as an example of the general approach. There are a great number of other charts (and other statistical tools) that are available to help improve the quality of the output of your process. The Xbar chart is a commonly used chart to describe the behavior of the center of your process - and process center is one of the key issues in creating high quality output.

YOUR PROCESS SHOULD BE CENTERED ON NOMINAL! If you think not, please consult your local (friendly) shrink.

In the following paragraphs I will discuss some of the more commonly used charts and, I might add, some of the less commonly used charts. I will try to give you some ideas as to the situations where each of the charts might be useful. I might also add that there may be some differences in opinion amongst those who read my comments, but the rational people will agree with me - as they always (never, sometimes, often, seldom, etc.) do.

VARIABLES CHARTS:

First off, charts in this category deal with properties that you measure with some sort of measuring device - not those that result from counting something or other.

There are usually two properties that we try to control for any variable being studied: the process center, and the process width. A chart for controlling the whole curve shape rather than just several characteristics will be discussed later in this section.

The process CENTER is controlled by looking at:

- The individual X's - see the discussion above concerning difficulties.
- The subgroup averages
- The sum of the deviations of the X's from the overall average
- Variations of the above - such as moving averages

The process WIDTH is controlled by looking at:

- The range
- The standard deviation
- Variations of these two - such as the moving range

Variables charts are usually considered in pairs such as Xbar and Sigma, or Xbar and R (for range), or X and Moving R, or Moving Xbar and Moving R.

Range charts are generally less efficient than sigma charts - statistical theory formalizes this idea. The range uses only two data points out of the sample. The standard deviation uses all of the data. Thus, for subgroups of size two, the range and the standard deviation have the same information content. Range charts have been made obsolete with the advent of the inexpensive high powered computers. But they are still in use, so they are still included in most SPC packages - even ours - since the customer is monarch.

Variables Charts for Process Center

X Charts

: (Individuals, Run, X - Moving R)

Purpose	:	Check to see if the center of the process is stable
Type of data	:	Variable
Centerline	:	Average of all of the X's - Xbar
Upper Limit	:	Xbar + 3 * Standard Deviation
Lower Limit	:	Xbar - 3 * Standard Deviation
Primary Use	:	Situations where data is accumulated slowly
Remarks	:	The limitations of X charts are discussed above - non-Normality of the process alters risks significantly.
	:	These charts are sometimes referred to as "run" charts.
	:	The major advantage is that Spec Limits as well as upper and lower control limits can be shown jointly on this chart
	:	There are many possible "standard deviations" that can be used in the limits:

The one calculated from all of the Xs - the major difficulty here is that the charts will tend to maintain the status quo rather than improve quality.

. One of the three possibilities mentioned below in the discussion of R charts, Sigma charts and S charts - the major problem here is that subgroups must be formed to obtain standard deviation estimates.

. The square root of the mean square successive differences estimate mentioned.

. The "Moving Range" - see WIDTH charts below.

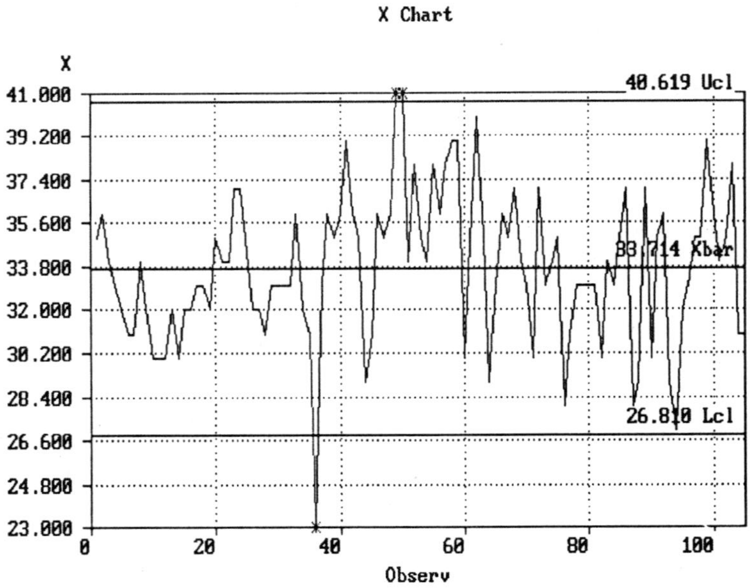

Figure 7

The chart in Figure 7 is an X chart for the same data as in the Xbar chart above. A subgroup size of three was used to develop the limits.

CumSum Charts:

"What is one?" you ask. Again I will resort to an example rather than a long definition. The data to be considered is as follows - in the order in which it was collected: 3,3,4,4,5,5. Deviations from the average or from the nominal value are usually used so as to keep the chart on a reasonable scale. In the example below I will use deviations from the average value.

N	X	X-Xbar	Sum of X-Xbar
1	3	-1	-1
2	3	-1	-2
3	4	0	-2
4	4	0	-2
5	5	1	-1
6	5	1	0

A plot of the cumulative deviations is shown in Figure 8.

```
Cumulative   1   !
                 !
                 !
  Sum        0   !                   *
                 !
                 !
  of        -1   !   *               *
                 !
Deviations       !
            -2   !       *   *   *
                 !
                 !---!---!---!---!---!---!---!---!---!
                 1   2   3   4   5   6   7   8   9

                          Observation Number
```

Fig.8

On Cumsum charts the downward trend indicates that the process is below average, the flat portion indicates that the process is right on average, and the upward trend indicates that the process is above average. The steeper the line slopes the larger the deviation from the process average during that time period.

Purpose	: Usually used to check process center for stability
Type of data	: Usually variable
Centerline	: None
Upper Limit	: Variable - either V shaped or parabolic shaped. Custom/QC uses parabolic "masks" for control limits.
Lower Limit	: Same as Upper Limit

Primary use : For low volume, slow data generation situations - the same situations for which X-Moving R charts are prescribed.

Remarks : These charts have the advantages of both the Xbar and the X charts:

- Like the Xbar charts they are insensitive to the shape of the process distribution.
- Like the X charts they allow plotting without waiting for enough data to make an average value.

: The chart is so visually powerful that the control limits are almost not required

: There are several disadvantages to these charts:

- They are new (1950s).
- The vertical scale does not allow direct readout of the process center. We have designed a nomograph to help in overcoming this problem - let us know if you would like a writeup about it.

See Appendix 3 for the average run length table for our CumSum parabolic mask.

Figure 9 is the data used in the charts above - in a CumSum chart.

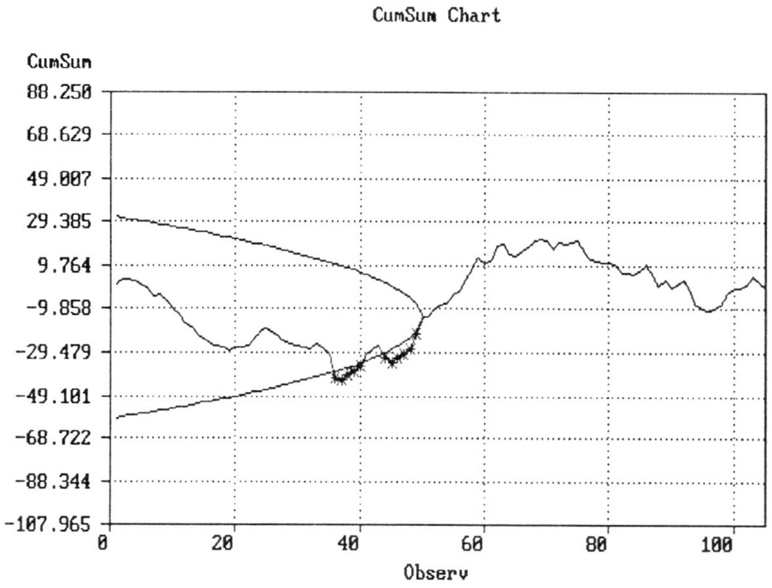

Figure 9

Min-Max Charts:

(See A.J. Duncan(5), pages 379-381 for discussion and values for the factor A6)

Purpose : Check to see if center or width of process has changed

Type of data : Variable

Centerline : Xbarbar

Upper Limit : Xbarbar + (A6 * Rbar)

Lower Limit : Xbarbar - (A6 * Rbar)

Primary Use : An alternative to the Xbar and R charts

Remarks : These charts are fundamentally X charts, but for each subgroup you plot the max and the min of the subgroup.

: The advantage of this type of chart is that both the width and the center of the process are being shown on a single chart.

: Research indicates that these charts are, for Normal processes, of about the same strength as Xbar and R.

: The major disadvantage is again the problems associated with frequency curves that are not Normal.

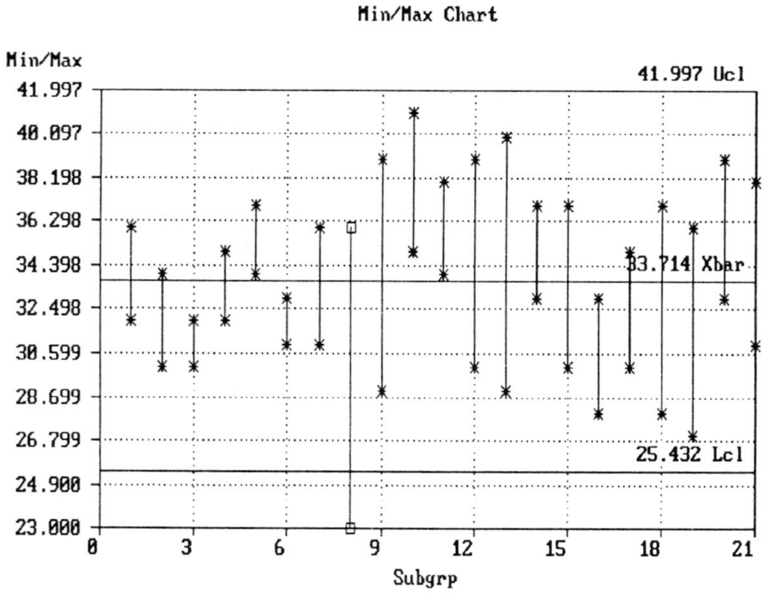

Figure 10

Xbar Charts:

Purpose	:	Check to see if center of process is stable
Type of data	:	Variable
Centerline	:	Average of subgroup averages - Xbarbar
Upper Limit	:	Xbarbar + A2*Rbar (or Xbarbar + A1*Sigmabar)
Lower Limit	:	Xbarbar - A2*Rbar (or Xbarbar - A1*Sigmabar)
Primary Use	:	Situations of medium to high volume production where data is quickly available for control purposes
Remarks	:	This one, and its companion R chart, are the charts that were first designed and used by Walter Shewhart.
	:	The Xbar chart is insensitive to the basic shape of the the process distribution. The shape of the Xbar distribution is approximately Normal under most conditions. This fact is referred to as the "Central Limit Theorem". Only severely skewed curves require fairly large sample sizes for this to hold.
	:	The values for Rbar and Sigmabar are discussed below in the section dealing with charts for process width.

I will not put another copy of the xbar chart here. I think you got the idea the first time.

Variables Charts for Process Width

Range Charts:

Purpose	:	Check to see if the width of the process is stable
Type of data	:	Variable
Centerline	:	Average of the subgroup ranges - Rbar
Upper Limit	:	D4 * Rbar (D3 and D4 are found in Appendix A.)
Lower Limit	:	D3 * Rbar
Remarks	:	Although the distribution of Ranges is not Normal for small samples, we use the same decision rules as for Averages. When we do this, we accept the fact that the risks associated with using the rules are just approximately the same as the risks when using Xbar charts. The distribution for ranges is skewed to the right. Thus, runs below the average will be more frequent on range charts than on the Xbar charts.

: The Range chart should be in control for the Xbar chart signals to have a clear meaning.

: Dividing Rbar by the factor d2 (see Appendix A) produces an estimate of the process standard deviation. This estimate multiplied by 6 is a measure of process capability

. The Range chart should be in control prior to using the term "capability".

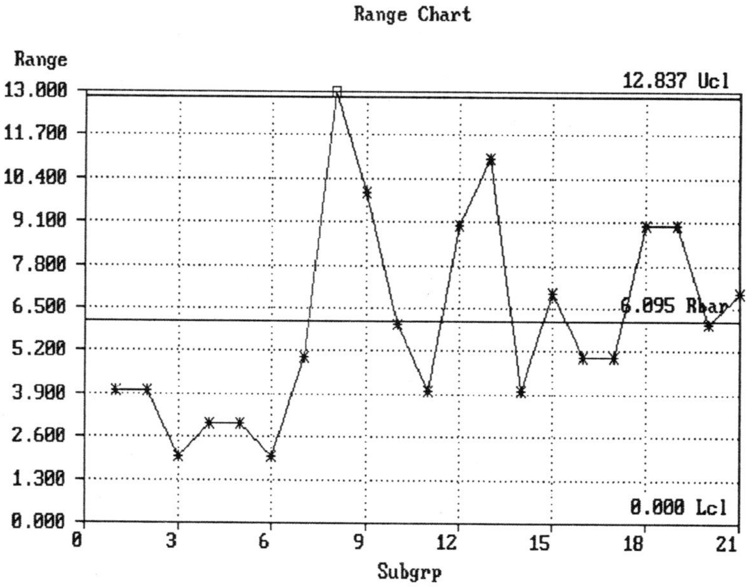

Figure 11

Sigma Charts:

Purpose	:	Check to see if the width of the process is stable
Type of data	:	Variable
Centerline	:	Average of subgroup standard deviations (using n - Sigmabar)
Upper Limit	:	B4 * Sigmabar (B3 and B4 are found in Appendix A)
Lower Limit	:	B3 * Sigmabar
Remarks	:	Generally the same as for R charts except that the process capability is calculated by using Sigmabar divided by c2 rather than d2 (see Appendix A for values of c2 and d2)

Figure 12 is the sigma chart for the same data that has been used for all of the variables charts.

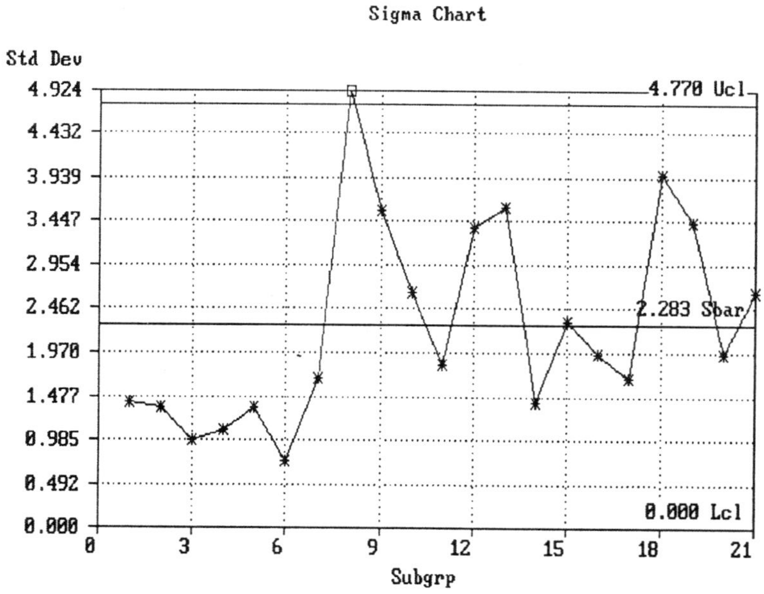

Figure 12

S Charts:

Remarks : Generally the same as for Sigma charts. The difference is that the standard deviation is calculated using n-1 rather than n. This in turn means that the factors B4 and B3 must be modified prior to use with the S chart. There is no particular advantage of either S or Sigma over each other. However, both S and Sigma have more power to detect process changes than does R.

: This chart was probably designed to overcome the sampling error problem mentioned in the discussion of standard deviations. The control chart factors make this correction and hence the S chart is superfluous - in my estimation. The "bias" in the standard deviation estimator remains, albeit to a lesser degree. Known bias can be accommodated in either case. However, there is a move underway to make all statistical definitions consistent - and the s chart is a step in that direction.

Moving R Charts : This is a concept that hasn't been addressed before. So, a definition is in order. Suppose the data is: 3,2,5,4,6,7 - in the order of occurrence.

N	X	Moving R	
1	3		
2	2	3 - 2 = 1	the range of the two numbers 3,2
3	5	5 - 2 = 3	the range of the two numbers 2,5
4	4	5 - 4 = 1	the range of the two numbers 5,4
5	6	6 - 4 = 2	the range of the two numbers 4,6
6	7	7 - 6 = 1	the range of the two numbers 6,7

Purpose : Check to see if process width is stable

Type of data : Variable

Centerline : Average of the moving ranges - Rbar (in example Rbar = 8/5)

Upper Limit : D4 * Rbar (see Appendix A for values of D3 and D4)

Lower Limit : D3 * Rbar

Primary use : Situations where data is accumulated slowly and control must be exercised as data is collected rather than wait for enough data to make a standard Xbar and R chart

Remarks : Factors D4 and D3 are based on a subgroup of size 2.

: Conceptually the same idea as the successive difference estimate of sigma. Thus dividing the Rbar by d2 produces a standard deviation estimate that is approximately the minimum possible value for the existing process.

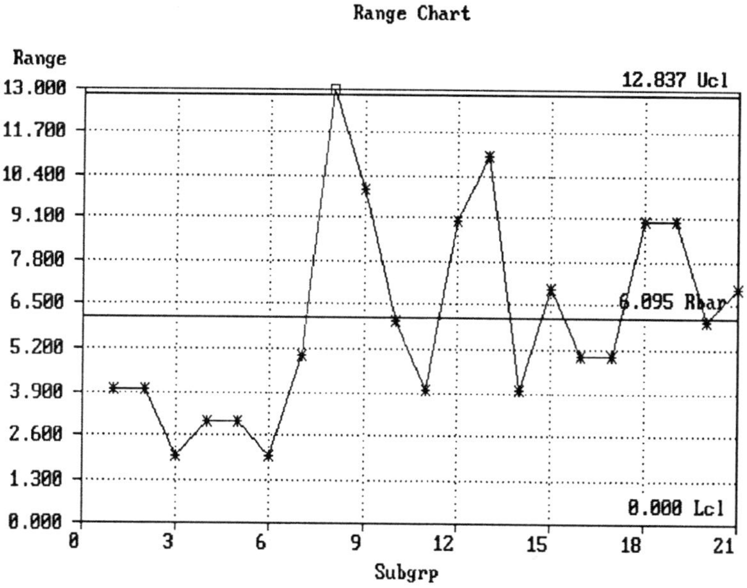

Figure 13

Z Chart:

I almost forgot this chart - the Z chart. First off, you do remember how to calculate Z of course. But let me refresh those who might have, in their enthusiasm to learn more, skipped that material.

Z(of any quantity Q) = (Q - average of Q)
/Standard Deviation of Q

A commonly used Z is the Z for Xbars. When you use "3 sigma" limits for Xbar charts, you are setting Z = 3. The calculation is done as follows:

Z(of Xbar) = (Xbar - Average of Xs)/
(Sigma of Xs / Sqr(n))

It doesn't look quite like the formula above, but that's just because I used two other ideas:

- The average of the Xbars is the same as the average of the Xs.
- The standard deviation of the Xbars is the standard deviation of the Xs divided by Sqr(n).

Purpose	: To check the center of a process for stability where a variety of products are being run through the same process
Type of data	: Quite general
Centerline	: Zero
Upper Limit	: 3 - if you are a "3 sigma" person (etc.)
Lower Limit	: -3 ditto
Remarks	: I'm going to be more wordy here, so I won't indent. I hope that suits you.

One of the nice properties of the Zs is that they are dimensionless and thus allow us to compare different products being manufactured on the same piece of equipment. In a "job" machine shop for example, parts with many different diameters which have in turn different averages and different sigmas, are sent through the same machine tool. Or in an ice cream plant, many different flavors with their different specs are sent through the same mixer.

The question is often asked, "How do I control such a process?". One approach is to plot, on the same chart, successive values of Z for the subgroups.

To do this, you need the average (which might be the nominal) and the sigma (from Process Capability Studies) for each product. Plug these values into the Z formula above to obtain the value of Z to be plotted. Notice that this Z formula also allows for changes in sample sizes that might occur amongst the subgroups. Custom/QC includes a Z-Xbar chart as a part of its list of computer charts. Any points with Z outside +3 or -3 should be considered as an indication that the process has changed.

Variables Chart for Curve Shape–Chi-Square Chart:

Purpose	:	Check to see if the shape of the process curve is stable
Type of data	:	Variable data which has been made into histogram form
Centerline	:	Number of classes - 1
Upper Limit	:	Center + 3*Sqr(2*(Number of classes - 1))
Lower Limit	:	Center - 3*Sqr(2*(Number of classes - 1))
Remarks	:	This type of chart was another one advocated by Shewhart(2) in his text. He said that while you can get some quality improvement using Xbar and R charts, the real question is, Did the shape of the curve change? And, he suggested the use of Chi-Square charts for that purpose. The reason that these charts have not been implemented to date is that they are a royal pain to do manually. But now that the computer is here, let's put it to work to do the things that need to be done. We have published an article on these charts in the Quality Magazine, February 1986, but I will describe them briefly here.

The idea is to generate the histograms for a series of subgroups; find the average histogram; and compare subgroup histograms with the average one.

The measure, Chi-Square, is one which summarizes the differences between the observed frequencies and the average frequencies. An example should do the job of describing the approach.

Class	Ave Freq	Obs Freq	Ave-Obs Freq	(Ave-Obs)^2	(Ave-Obs)^2/Ave
1	20	25	-5	25	25/20 = 1.25
2	30	25	+5	25	25/30 = 0.83
3	40	42	-2	4	4/40 = 0.10
4	20	24	-4	16	16/20 = 0.80
5	10	4	+6	36	36/10 = 3.60

					Chi-Square = 6.58

Now the next things you need to know, for control chart purposes, are the values for the centerline and the control limits. This part is easy. The centerline - the average value for Chi-Square - is equal to the number of classes minus one. (In this case 5 - 1 = 4). T/e sigma value for Chi-Square is Sqr(2*(k-1)). (In this case Sqr(8) = 2.828). Thus, the three sigma limits are established as being:

UCL = 4 + 3 * 2.828 = 12.84
LCL = 4 - 3 * 2.828 = 0
(Since the value of Chi-Square cannot be negative)

The value of 6.58 lies inside these limits and hence you would conclude that the observed curve is not a whole lot different from the average curve.

The Chi-Square curve is not symmetric, so using the 3 sigma limits is just an approximation to what you would find if you were to use the table value for Chi-Square - in this case approximately 13.5 as opposed to the 12.8. The Chi-Square tables require "degrees of freedom" to find the probabilities. When using the Chi-Square chart in this fashion, the number of degrees of freedom is the number of classes minus one. So if you want to be more exact, just use a Chi-Square table from Custom/QC or the text of your choice. (Duncan(6) talks about using the Chi-Square chart for a set of fraction defectives.)

Since the Quality article has been published, a graduate student named Larry Gioia(7) has done some simulation runs to check the effectiveness of the Chi-Square charts under less stringent process shifts than those used in the article. He generally found them to be as good or better than the Xbar - R charts. The better part came when both the average and standard deviation of the process had changed at the same time.

He pointed out a disadvantage of the charts: it is not clear from the chart alone what the characteristics were of the process shift. This can very readily be remedied by comparing the histogram of the data prior to the change with the histogram of the data after the change. Or, alternatively, by comparing the mean, standard deviation, skewness and kurtosis of the subgroups from which the chart was built.

THE LAST OF THE VARIABLES CHARTS - in this text. I know you have been looking forward to this section.

The world of manufacturing can be divided into two camps: the cooks and the whittlers. Of course some of you do both. Some of you cook and then whittle; some whittle and then cook. This section is of most interest to the cooks. But I have seen the chart described below used by whittlers on occasion.

"But", you cooks say, "I have two or more, possibly correlated, variables to control. What can I do with all of that data?". One possibility is to make individual charts for each variable. That can be so time consuming that there is no time left over for production. Another problem is that the risks you thought you were taking are eroded roughly in direct proportion to the number of charts you use. And as if these two problems are not enough, the existence of correlation amongst the variables can cause changes to be overlooked. For a good writeup of this situation please refer to Jackson's article in the Journal of Quality Technology, October 1980. In that article he discusses the construction and use of T-Square control charts. They are also referred to as Hotelling's T-Square charts in honor of their originator.

Multi Variable Control Charts–T-Square:

Purpose	: Detect process shifts from a set of two or more, possibly correlated variables
Type of data	: Variable
Centerline and Limits	: I really shouldn't bother you with these in this discussion. It is enough to say that there is a center line and an upper control limit. The calculations are complex and best handled by the computer.

Remarks

: These charts, like the Cumsum and Chi-Square charts, have been just waiting for inexpensive computers. The calculations are roughly equivalent to doing a multiple regression study on each subgroup. Those of you who don't know about multiple regression please see our Volume II (later) or consult your friendly statistician (YFS). The remainder of you understand why these charts are not in the top 10 of the control chart hit list.

: These charts are rather difficult to interpret - a point out-of-control is a signal that one or more of the averages have changed or that the usual relationship amongst the variables has changed. The question is "Which of the above and by how much?". A possible solution is to move all of the variables onto a single (principal) axis. Once the variables have been moved to this axis, the culprit can be identified. But, here again, you may need help from YFS.

: There are some cases where the interpretation is not as difficult as just suggested. An example of such a use is a T-Square chart being used on screen analysis data collected to determine particle size distribution changes from one lot of raw material to the next. A point out-of-control in this case means that the distribution has changed and the culprit is usually quite obvious. An aside - a Chi-Square chart would be equally applicable for the example if the count of particles rather than the percent by weight were available. Please call us if you would like a copy of a paper delivered to the American Ceramics Society in May 1986 dealing with this topic. We presented a paper on this topic to the Iron and Steel Institute in Oct 1987 - different data and showed how to identify the key variables on which to run the chart.

The T square chart for the Iron and Steel paper is shown below in Figure 14.

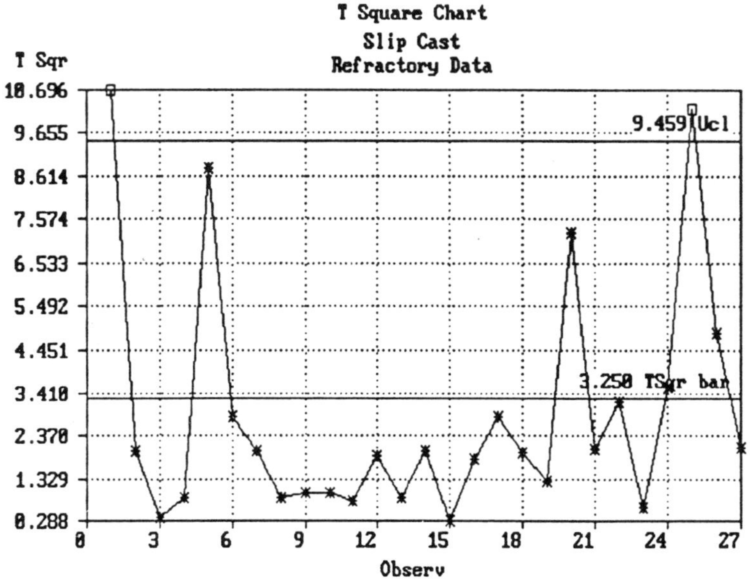

Figure 14

Next we need to discuss ATTRIBUTE CHARTS.

Defectives Charts:

The most commonly used chart in this category is one for the fraction defective of a process. The "fraction defective" is obtained by dividing the number of bad items in a sample (or universe) by the number of items in the sample (or universe).

When the fraction defective is multiplied by 100, it is referred to as the percent defective. The chart for fraction defective is referred to as a "P Chart".

Charts in this general category are appropriate for process control if there is no measurement data available - such as ball point pens either writing properly or not writing properly. Or, it may be that measurements are too expensive. They are often used to summarize the quality level of an operation on a daily, weekly, or monthly basis.

This class of charts is interesting because a point out-of-control on the low side is an indication of significant process improvement. That's why I used the expression "capitalize on the out-of-control information."

P Charts:

Purpose	: To check the process percent defective for stability
Type of data	: Attribute
Centerline	: Average of subgroup P values - Pbar
Upper Limit	: Pbar + 3*Sqr(Pbar*(1 - Pbar) / n)
Lower Limit	: Pbar - 3*Sqr(Pbar*(1 - Pbar) / n)
Primary use	: For situations where measurement data is not available and sample sizes vary. Used for both process control and management summary purposes.
Remarks	: The width of the control limits change as n changes which makes the chart a little difficult to interpret.
	: If you are monitoring the percent defective of a set of characteristics rather than a single one, use the Chi-Square chart discussed above.

A sample of a P chart is shown below.

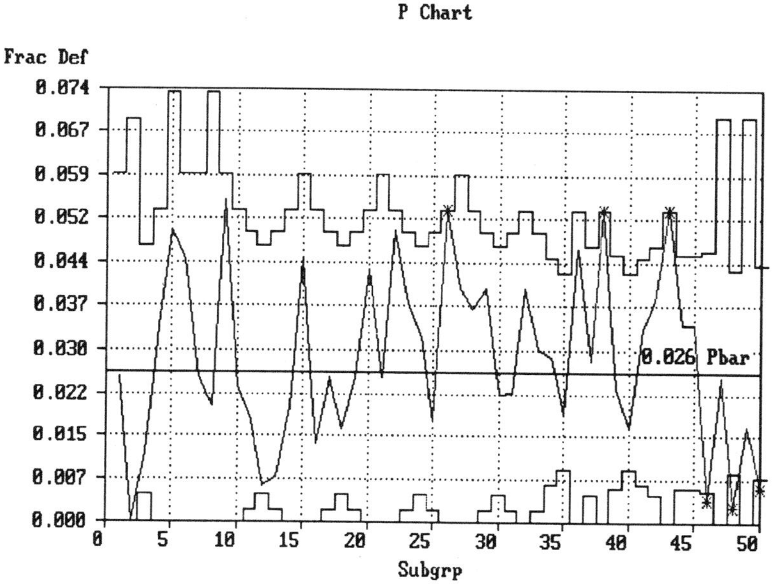

Figure 15

NP Charts:

Purpose	: Similar to the P chart except that it deals with the number bad rather than the percent bad. It is used only where the sample sizes of each subgroup are the same.
Type of data	: Attribute
Centerline	: n * Pbar
Upper Limit	: n*Pbar + 3*Sqr(n*Pbar*(1 - Pbar))
Lower Limit	: n*Pbar - 3*Sqr(n*Pbar*(1 - Pbar))
Primary use	: See P Chart discussion above

Remarks : Most likely to show up in process control situations where the sample size is fixed.

: Easier to plot and understand since control limits are of constant width.

I'll put an NP chart in here even though I suspect it's a little overkill - you'd think I'm getting paid by the page.

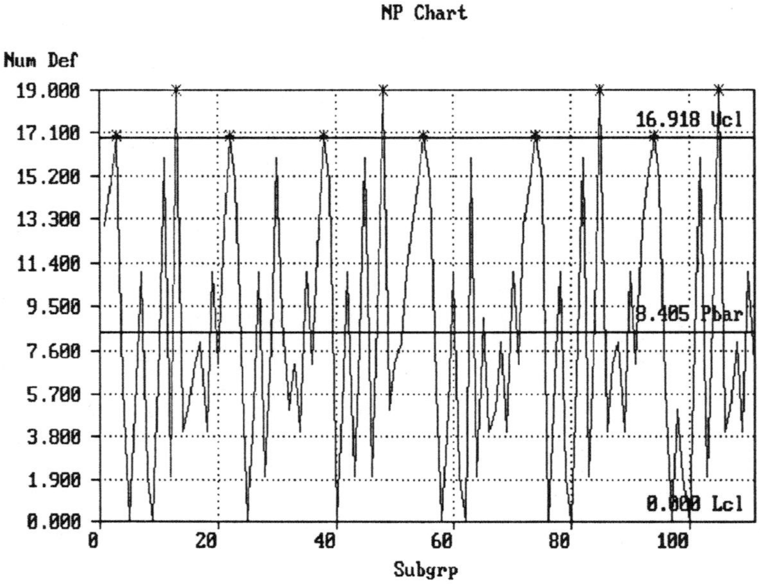

Figure 16

The P chart and the NP chart are used for controlling DEFECTIVES. The next chart pair is used for controlling DEFECTS. I think that you know the difference between DEFECTS and DEFECTIVES - it may take several defects on a part before the part is deemed to be defective.

Defects Charts:

C Charts:

Purpose	: Check to see that number of defects being generated by the process is stable
Type of data	: Attribute
Centerline	: The average of subgroup numbers of defects - Cbar
Upper Limit	: Cbar + 3*Sqr(Cbar)
Lower Limit	: Cbar - 3*Sqr(Cbar)
Primary use	: Analysis of customer complaints, number of flaws per square yard of material and so on
Remarks	: This chart is similar to the NP chart in that the area of opportunity must be the same from subgroup to subgroup. That is, the number of defects must be observed on a constant size of cloth sample.

Now I could put a C chart in here but it would look just like the NP chart above. So, I'll spare you that.

U Charts:

Purpose	: This chart is the defects parallel to the P Chart in that the size of the opportunity of an occurrence changes from subgroup to subgroup and this must be allowed for in the calculation of the control limits.
Definition	: U = Number of occurrences / Area of opportunity
Primary use	: Situations such as Number of Accidents / Man Hour
Centerline	: Total Number of Occurrences in the subgroups divided by the total area of opportunity in the subgroups : Ubar
Upper Limit	: Ubar + 3*Sqr(Ubar/Area of Opportunity)
Lower Limit	: Ubar - 3*Sqr(Ubar/Area of Opportunity)
Remarks	: As indicated in the case of P charts, the control limits vary as the area of opportunity for the defects to occur changes.

And, of course a U chart would be very similar in form to the P chart. So, I'll skip that one too.

If you find that the charts I skipped are the ones you really wanted to see, give us a call and we'll see that you get some examples.

This is the end of our discussion of types of control charts. We now move on to the question of subgroup or sample size.

Selecting the Appropriate Subgroup Size– Related to "Rational Subgrouping"

I guess this section should be subtitled : Is your control chart telling you what you need to know? And maybe I'd better give an example of several which did not so that you can see the purpose of all this.

For the cooks amongst us (Whittlers don't despair! There's one for you too).

Samples of size 5 were being taken from each batch of product being generated. Since subgroup size 5 is the most magical number, control charts were built using a subgroup size 5. The control chart showed about 2 out of 3 averages outside the control limits - in a fairly random fashion. The explanation of this phenomenon: "There is more variability between batches than one would expect given the variability within batches". Now you cooks don't need a control chart or a fancy statistician to tell you that. That's a fact of life.

It's also a fact that reducing the batch to batch variability to the same level as the within batch variability is a worthwhile GOAL. But we move toward that goal by finding and eliminating unusual, as opposed to facts of life, events. Try using a subgroup size 15 (three batches worth of data). This allows for what Gerry Hahn(8) refers to as "legitimate batch to batch variability". Points outside the control limits now indicate that a set of three batches has something unusual about it when compared to other sets of three batches. Find and eliminate that source of variability and you will move along the way toward your goal. Use the control charts as a tactical weapon which will help you achieve your goal. Move from being able to control at the "three batch level" to being able to control at the "two batch level".

For you whittlers: please refer to the example above.

Oh, all right, I'll give it a try. A very nice example of this sort is discussed in Grant & Leavenworth(2), pages 160-163. I will summarize it quickly.

A semi-circular engine mount has four pads on its inner surface. Data on the thickness of these pads is collected for 36 mounts. Control charts are made using subgroup size four - of course. Grant & Leavenworth say " --- it was practically certain this type of subgrouping would show lack of control, and that this showing lack of control did not provide a useful guide to action. --- That is, it was to be expected from the fact that the four pads on any half-ring were machined together; it was reasonable to expect more variation among the half-rings than among the pads. Moreover, it was of no help to know that this was true, as nothing could be done about it by attention to out-of-control points."

Grant and Leavenworth's approach was to make a control chart on each pad, separately, using a subgroup which covered multiple half-rings. Note that a T-Square chart might also be a useful tool in this situation - avoid making charts for each pad until the T-Square on the four pads indicates lack of control.

Another approach would be to take a subgroup of size twelve (three half-rings) in order to allow for some legitimate ring to ring variability to be included in the control limit calculation.

To summarize, we have to go back to the question of:

What is the purpose of an SPC chart?

Answer: To alert people to unusual events so that causes can be identified and appropriate action taken to improve quality.

Custom/QC has, in its " Process Capability Studies" module, a sweep over a large number of subgroup sizes and indicates whether or not the process would be in control for each size. This approach allows you to pick a subgroup size which would help you move along the path of improved quality.

This sweep, on Grant and Leavenworth's(9) Example 5.2, indicates that 3 or 4 half-rings would make a subgroup size for which the process would be in control most of the time. Using this subgroup size, we can arrive at stable estimates of the process capability. If the stable process is producing poor quality product, then the subgroup size should be altered in such a fashion as to provide occasional out-of-control signals. If the stable process is producing acceptable quality product, then maintain control at this level for this variable and move on to investigating and controlling other process variables.

A graph of the sweep is shown in Figure 17. Subgroup sizes for which the probability on the vertical axis is about .98 or less will likely yield an Xbar chart which is in control.

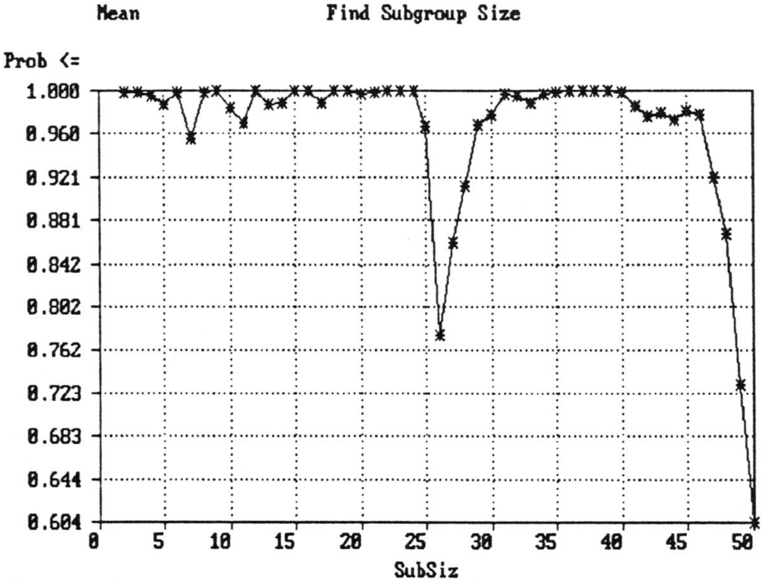

Figure 17

From this chart you can see that a subgroup size 10 will be in control. The subgroup size 10 chart is shown below in Figure 18.

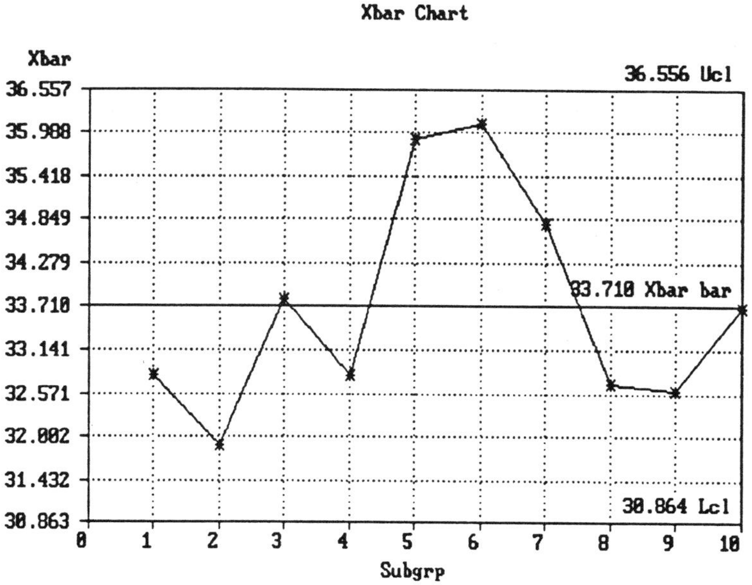

Figure 18

There are two countervailing forces at work as the subgroup size increases:

. The within group variability increases - making the chart less sensitive to changes in process average values (or widths). The Average Run Length (ARL) increases.

. The standard deviation of the averages decreases, due to the larger sample size - making the chart more sensitive to changes in process average values (or widths). The ARL decreases.

So you can see that the use of larger subgroup sizes is not, in itself, bad, and may well result in the constructive use of control charts in your plant.

There are other approaches to this question of subgroup size.

Duncan and others have constructed cost models which, given some cost factors, enable you to find the subgroup size and frequency. The total cost is a function of the Average Run Length, the cost of taking samples, the cost of searching for trouble, and the cost of being out-of-control. The ARL is a function of the subgroup size, so one can arrive at some reasonable estimates of subgroup size from the model. However, I should add that the model assumes that the subgroup so selected is "rational" - in the context of the examples above. In summary, awareness on your part of the concepts of ARL and rational subgrouping will take you a long way toward selecting the proper subgroup sizes for your charts.

FINAL GASP ON CONTROL CHARTS USED FOR SPC PURPOSES

I have now finished two major sections of this text: Descriptive Statistics and Control Charts. "Why," you ask, "bother with the distinction?" Should you be asking that question, let me assure you that you are not alone.

Descriptive Statistics applied to quality data provide a managerial summary of the results of all of the quality efforts during the day, week, or month. They are static in nature.

Control Charts used for SPC purposes are dynamic in nature. They deal with controlling the process so that the descriptive statistics show improvement. They are used to control the process with a level of detail that is usually not of importance to the manager - he or she does not generally need to know the results of each subgroup. He or she needs to know that appropriate action is being taken and that the results as indicated by the descriptive statistics summary are good and getting better.

The financial accounting community has this situation pretty well under control. They have the Balance Sheet (equivalent to our Descriptive Statistics) and the Income Statement (equivalent to our Control Charts). A balance sheet that shows an excellent cash position today (static) is good - but what if the income statement shows that it has been dropping drastically (dynamic)? The folks in charge of getting good product to the customers should not be limited to a Balance Sheet - USE CONTROL CHARTS for the dynamics of process control!

PROCESS CAPABILITY - one last time

When the control charts for the average and the standard deviation indicate control, then you can rely on the value of Xbarbar and Sigmabar as estimates of the "true " process values. From these true values you can approximate the percentage of your product that will meet specifications. And you can calculate the values for the CP, CR, CPK and so on. These values are DESCRIPTIVE measures of a process which is in control. Some folks think of them as process control devices. They are not - at least in the manner they are usually used. There are ways to put control limits on the CPs and the CPKs but this analysis is basically just setting up control limits on 1/Sigma rather than Sigma. My reaction here is similar to my reaction to the "s" versus the "sigma" chart - they both work so pick the one that is most satisfying to you.

The "sweep" method that I just discussed is one way to obtain estimates of the process capability from a "constant system of chance causes" - which Shewhart referred to as a controlled process.

Another procedure that is used is to accept the fixed (God given?) subgroup size. Then out-of-control subgroups are eliminated until finally all points are in control. At that point the process capability calculations are invoked. These measures are based on the assumption that all of the causes for the out-of-control points have been identified and eliminated. This just might be tough to accomplish - but it is another approach.

Sample Size for Estimate of Process Average

When we use control charts, we use many small samples over time (usually) to tell us whether or not the process has changed. A point outside the 3 Sigma limits on an Xbar chart is an indication that the process average has changed. You probably would be disinclined to bet your next two weeks wages that the process average is the same as the average of the small subgroup.

So, let's look at another question: How large a sample size do you need to get a good estimate of the process average? (We'll look at the same question relative to standard deviation and to fraction defective a little later.) Please note the difference in the questions being asked:

- Did the process average change? (Control Chart)
- What is the process average? (Sample Size Calculation)

To obtain the answer to the latter question, we have to state:

- The maximum error size that we can tolerate - the smaller the error, the larger the sample size
- The risk we are willing to take that the maximum error will not be exceeded - the smaller the risk, the larger the sample size. (I might as well add that the larger the Sigma of Xs, the larger the sample size.)

For example, suppose we would like to know the average value of the process with a maximum error of 0.02 and that we would like to be about 99.7% sure that the error (E) made will not exceed that value (0.02). Assume too that we know that the process Sigma is 0.1 from previous studies. (Call your friendly statistician if you don't know Sigma. He or she will know how to solve this problem - but it detracts from our discussion of fundamentals.) The solution is obtained by using three facts that have been mentioned before:

- The distribution of averages is approximately Normal: Z = 3 for this example to get the 99.7% assurance specified.

- The average of the Xbars is the same as the average of the process from which the samples are drawn: E = Xbar - Process Average = .02 in this example.

- Sigma of the Xbars is Sigma of the Xs divided by Sqr(n): Sigma of the Xbars is 0.02 / Sqr(n) in this example.

The answer is obtained by solving the equation:

3 = 0.02 / 0.1/ Sqr(n)

for n. In this case n = 9 * .0100 / .0004 = 225.

In general, to get the sample size for estimating the process average use:

n = (Sigma of Xs * Z / E) squared =
(0.10 * 3 / 0.02) squared = 225.

Sample Size to Estimate Process Sigma

The sample size required to estimate Sigma of the Xs is best obtained from tables such as Dixon and Massey Table A-6c (1). I don't think I'll bother with an example here. Just refer to D&M if you have a problem of this sort.

SAMPLE SIZE FOR ESTIMATE OF PROCESS FRACTION DEFECTIVE

The sample size required to estimate the process fraction defective is found (based on the same approach used for the average above) by:

n = (Z / 2E) squared

Thus if you wanted to estimate the fraction defective of a process with an error (absolute) of less than 0.02 with 99.7% assurance, the sample size would be:

n = (3 / 0.04) squared = 75*75 = 5625

Oh, that's a big sample size! The magnitude of the difference of the sample size for Xbar and the sample size for fraction defective is one of the arguments for using Xbar charts rather than P charts wherever possible.

Acceptance Sampling

I included the sample size discussion above as a lead in for our next topic: Acceptance Sampling. The most widely used documents covering this material are the two US military standards known as:

Mil Std 105D	for attribute sampling
Mil Std 414	for variables sampling

The basic question being asked here is: Based on sample results, should I Accept or Reject the lot of material that I just received from my supplier? The standards referred to contain sets of sampling plans which are used to answer this lot disposition question.

The set of questions we have addressed recently is now expanded to:

- Did the process average change? (Control Chart)
- What is the process average? (Sample Size Calculation)
- Accept or Reject the lot based on sample results? (Acceptance Sampling)

THE RISKS OF ACCEPTANCE SAMPLING PLANS

Let's review the risks involved with Xbar charts first.

A. The Xbar chart indicates lack of control but the process average has not changed. In this case you will either:

- take action on the process unnecessarily and introduce extra variability in your product OR
- take a large sample (size determined as above) to find out what's going on - and find nothing interesting

If you are using just the 3 Sigma limits on your chart, then the chance of making this kind of mistake is about 3 in a thousand.

This type of risk is referred to as the "Alpha" risk - that's why I put an "A" on the front of this paragraph.

B. The process average value can change, but the Xbar chart does not indicate lack of control. In this case you take no action but the process goes on its merry way pumping out bad stuff.

This type of risk is referred to as the "Beta" risk - I suspect that you know why I put a "B" in front of this paragraph.

Now let's develop the parallels of these risks in the context of Lot Acceptance Sampling Plans using attribute data.

The first thing we have to do is to separate - once more - GOALS from TACTICS.

ZERO DEFECTS is a good goal. It's one helluva tactic - to be sure that the lot you are inspecting contains no bad items you must resort to about 300% inspection. Now, there's nothing wrong with that as a tactic - given an overall objective appraisal of the relative costs. What is wrong is to allow the goal to interfere with how you think about the risks involved - once you have decided to use an Acceptance Sampling Plan.

!! THIS SECTION ASSUMES THAT YOU HAVE MADE THE DECISION TO USE SUCH A PLAN !!

There are two concepts that are helpful in the evaluation or determination of an Acceptance Sampling Plan: AQL and RQL.

AQL - The Acceptable Quality Level

Now I know that this term has fallen into ill repute (ZERO DEFECTS, PPM and all that) but this is purely a matter of semantics - not of evaluating sampling plan risks. The idea is that if the lot has a small fraction defective, then the lot should be "accepted" by the sampling plan most of the time. The lot fraction defective that should be accepted most of the time is referred to as the "Acceptable Quality Level" (AQL).

RQL - The Rejectable Quality Level

There is another idea that will help in our evaluation of a sampling plan - if a lot has a percent defective that exceeds some number, e.g. 5%, then the lot should be "rejected" by the sampling plan most of the time. This idea is referred to as the "Rejectable Quality Level" (RQL). Another term that is used in this context is the Lot Tolerance Percent Defective (LTPD).

Now let's draw up the risks for the sampling plan in a fashion similar to the approach used for control charts.

A. The lot fraction bad can be better than the Acceptable Quality Level (AQL) and the sample results indicate that the lot should be rejected. This risk is called the Alpha Risk for the plan - the lot met the AQL (or better) but was rejected. Since you supplied an "acceptable" lot that was rejected, this risk is also referred to as the "Producer's Risk".

B. The lot fraction bad can be worse than the Rejectable Quality Level (RQL) and the sample results indicate that the lot should be accepted. This risk is called the Beta Risk for the plan - the lot didn't meet the required quality level but it was accepted. Since the customer receives this bum lot, this risk is also referred to as the "Consumer's Risk".

These are the basic concepts. The ideal sampling plan would accept all lots with fraction bad less than the Acceptable Quality Level (AQL) and reject all lots with fraction bad greater than the Rejectable Quality Level (RQL). Sampling plans can be constructed which approximate - but never achieve - the ideal. Once the values for AQL with its corresponding Alpha Risk and for RQL with its Beta Risk are established, sampling plans can be constructed which approximately meet these conditions.

Variables Sampling Plans–I'll get to the more common Attribute Plans later.

I'll do Variables first since we recently made a sample size calculation - to estimate the process average, not to decide whether to accept or reject the lot.

Variables plans generally make the Accept - Reject decision based on one of the following:

- Percentage of material not meeting specs

- Average value not meeting specified average value

- Standard deviation not meeting specified standard deviation

I will concentrate mainly on the first case - and even then on just the One - Sided Specification case. And, I will discuss only the case where sigma is assumed to be known.

Such a sampling plan, for a low end spec limit only, is described as follows:

- N = Lot size - but the lot size is not a particularly significant element in this discussion

- n = Sample size

- XbarL = Decision point for sample average. If the sample Xbar is below XbarL then the lot is rejected.

Graphically the situation looks like this:

```
                      Take a sample of size n.

                                 !
   If Xbar is here, then         !   If Xbar is here, then
       REJECT the lot            !       ACCEPT  the lot
                                 !
--------!------------------------!-------------------------
       LSL                     XbarL
```

You probably have noted a philosophical, and a potential legal problem here -

> All of the sample values can be above the Lower Spec Limit and yet the sample average may be in the REJECT zone.

Brushing this problem aside in a cavalier manner, we move on to technical considerations.

The technical questions are: What value of n should be used to get the kind of protection that is required? And, what is the value for XbarL?

The answers to these questions require setting up and solving two equations rather than the single equation we used for sample size.

I'll show you the approach using an example:

Example:

Assume that sigma is 2 and that the Lower Spec Limit is 50.

Next assume that the Acceptable Quality Level (AQL) is .01 and that we would like to have lots of that quality accepted 95% of the time.

Finally assume that the Rejectable Quality Level (RQL) is .04 and that we would like to have lots of that quality accepted only 2% of the time.

These assumptions are summarized in the table below.

LSL	50	AQL	.01	RQL	.04
Sigma	2	Alpha	.05	Beta	.02

The process average which will produce 1% bad is 2.326 standard deviations above 50 - or in this case 54.652. If the process average is 54.652, then 95% of the Xbars will be above 54.652 minus 1.645 * 2/Sqr(n). So,

XbarL = 54.652 - 1.645 * 2/Sqr(n).

This is one of the two equations I mentioned.

The process average which will produce 4% bad is 1.751 standard deviations above 50 - or in this case 53.502. If the process average is 53.502, then 98% of the Xbar's will be above 53.502 plus 2.054 * 2/Sqr(n). So,

XbarL = 53.502 + 2.054 * 2/Sqr(n).

Now we just solve these two equations simultaneously for n and XbarL and we have the answer to the questions above.

The answers for this example are:

n = 42 and XbarL = 54.136.

In addition to the philosophical problem mentioned above, I should add that the risks involved with variables sampling plans are evaluated based on the assumption that the X's are Normally distributed. Any deviation from this assumption may have a sizable impact on the actual risks.

Well anyway, that's the approach that is used. Please see a text such as Duncan(5) for more details on the method for setting up these plans.

I can't leave this section without mentioning several "control charts" that are really acceptance sampling in disguise. The most notable of these is the "tool wear chart". This chart is made so that the lower (or upper) control limit is set relative to the lower (or upper) spec limit. The logic for determining the lower control limit is very similar to that mentioned above. The process average is obviously not stable in the tool wear case. The chart is being used to tell us when the parts being produced will be out of specs. Control could be tested if the control limits were established around the expected trend due to tool wear. This approach is discussed in the section dealing with regression. (Regression will be covered in the second volume of this book.)

This application is just a special case of "My spec limits are so wide that I can let the process slop around as long as the average value doesn't get too close to the spec limits!".

You may be contented with that situation. But, you should avoid using the term control chart when you are in effect doing acceptance sampling - albeit on the process not the lot. If you are contented with such an approach, then you have not taken Taguchi's message seriously!

Attribute Plans

An example of a single random attribute plan is:

Lot size (N)   = 1000
Sample Size (n)   = 50
Acceptance Number (c) = 2

This means that we will take a random sample of size 50 from a lot of size 1000. If we find 2 or fewer bad parts, we will accept the lot from which the sample was drawn. If we find more than 2 bad items in the sample, we will reject the lot.

An example of a double sampling plan is:

N = 1000
n1 = 50 c1 = 1
n2 = 100 c2 = 3

This one means take a random sample of size 50 from the lot of size 1000. If the number of bad items is 1 or fewer, accept the lot. If the number of bad items in the sample is 4 or more, reject the lot. If the number of bad items in the sample is 2 or 3, take a second sample of size 100. If the total number of bad items in both samples is 3 or fewer, then accept the lot. If the total number of bad items in both samples is 4 or more, then reject the lot.

Double sampling plans have several advantages:

. The risks can be moved more toward the ideal situation.

. The lot gets "a second chance."

Double sampling plans, however, require more managerial attention to ensure that they are properly used.

Multiple sampling just uses more than two samples in the Accept - Reject decision process. They get more complex to administer, but have some advantages along the lines of those mentioned for double sampling.

Sequential sampling looks at one item at a time. The decision is made on the basis of the total number of bad items found up until that time to either accept, reject, or continue sampling.

As we move from single to sequential, the plans get more complex to administer, but they also generally require smaller sample sizes to be used in reaching the Accept - Reject decision.

Suppose I used the AQL - RQL risks shown in the table below to construct a sampling plan.

Lot fraction defective		Chance of accepting
Acceptable Quality Level (AQL)	= .01	.95
Rejectable Quality Level (RQL)	= .05	.02

In fact, I used the "Sampling Plans" module of Custom/QC to construct just such a sampling plan - one which meets those criteria. The sampling plan is:

n = 218
c = 4

Notice that the solution does not refer to the Lot Size. That's because for large lots, the lot size is not particularly important and can be neglected in the calculations. I'll show you that a little later.

The Operating Characteristic Curve as a Description of Risk

"Yes, but how good is that sampling plan?", you ask. Let's see how the plan performs at various lot fraction defective levels, and then generalize the approach. To answer that question, let's start with the table above and add a couple of entries - that don't need any probability calculations.

If the lot has no bad items (0% defective) then any sampling plan will accept it - if you can't find a bad item, you sure can't reject the lot. So let's add this information to the table.

Lot fraction defective		Chance of accepting
	.00	1.00
Acceptable Quality Level	= .01	.95
Rejectable Quality Level	= .05	.02

If the lot contains all bad items (100% defective) then any sampling plan will reject the lot - there's no way to get even one good one in the sample. So let's add another line.

Lot fraction defective		Chance of accepting
.00		1.00
AQL = .01	approximately	.95
RQL = .05	approximately	.02
1.00		.00

The next thing to do is present the table in graphical form as shown below.

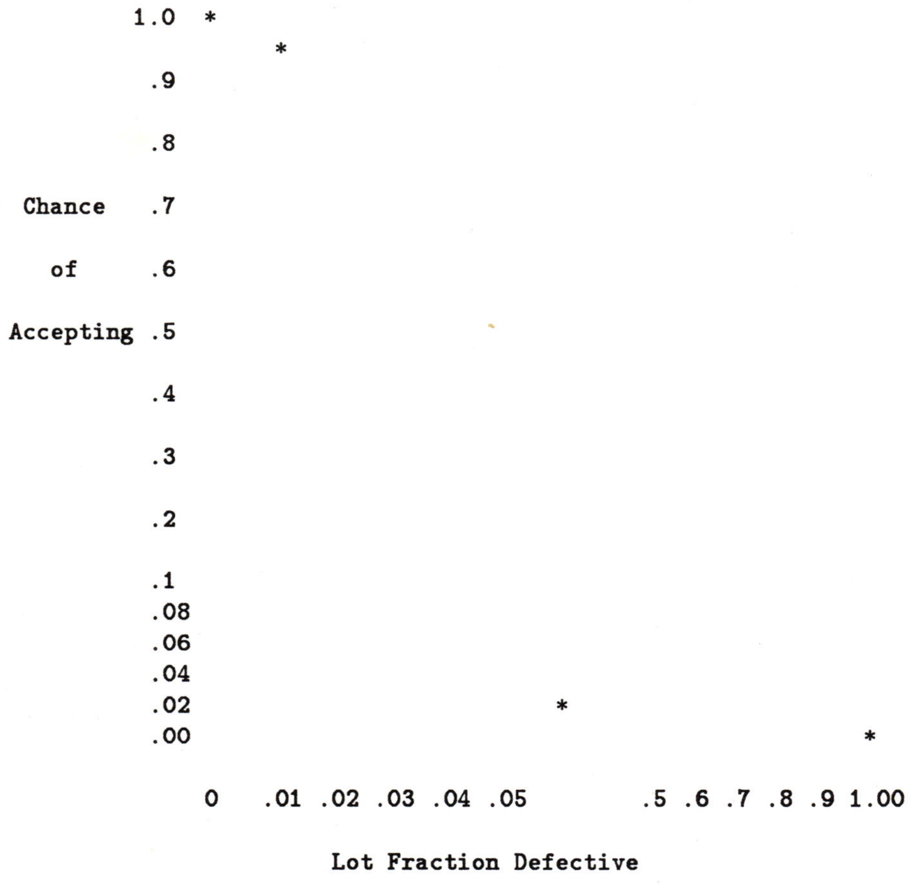

Figure 19

Notice that I have modified the scale on both the x axis and the y axis. I did that so that even with the small number of points, you can see a curve shaping up. The name of the curve is "The Operating Characteristic Curve". It is called "The OC Curve". Once you have the Operating Characteristic Curve for the sampling plan you know all there is to know about the plan's risks - assuming of course that your inspectors and gages are perfect.

At this point you really have to ask me to show you how to get more points on the Operating Characteristic (OC) Curve or else I don't get the chance to show you why the lot size is not particularly important.

Thank you for asking. Here comes some elementary probability!

How DO you get more of the points on the Operating Characteristic (OC) curve?

Let's see how to get more points on the OC curve for the sampling plan we have been using as an example: n = 218, c = 4, and for now assume that N = 1000.

The lot will be accepted if the number of bad items in the sample is 0, 1, 2, 3, or 4.

The chance that the sample contains 0 bad items is easy. Let's start there. In order for the sample to contain 0 bad items, each of the items in the sample must be a good one.

And, let's get the exact point on the OC curve corresponding to a lot whose fraction defective is .05. That means that the lot of 1000 contains 50 bad items and 950 good items. The calculations are shown in the table below.

Sample Item Number	1	2	3	4	5	6	. . .	215	216	217	218
Item Condition	G	G	G	G	G	G	. . .	G	G	G	G
Number Good Remaining in Lot	950	949	948	947	946	945	. . .	736	735	734	733
Number Items Remaining in Lot	1000	999	998	997	996	995	. . .	786	785	784	783
Prob	.9500	.9499	.9499	.9498	.9498	.9497	. . .	.9364	.9363	.9362	.9361

Remember, I said "easy" not "fast". Now all we have to do is multiply these 218 individual probabilities together to get the probability (chance) that the sample contains 0 bad items. I have to admit that I don't enjoy that kind of work, so I asked Custom/QC module "Frequency Curve Comparisons to do it for me - using the "Hypergeomtric Curve". The answer is 0.000003.

Before we go any further, notice that the probabilities are not too far from all being about .95. As a matter of fact, if I had used .95 rather than the exact probabilities, the answer would have been .000014 - the difference is way out in the fifth decimal point. Way out!

The fact that the .95 approximation to the actual probabilities gives roughly the same answer as the actual is the basis for the earlier statement that the lot size is not particularly significant. That is, we would have gotten approximately the same answer if we had used 218 of the ratios 950/1000 as we did with the ratios that took the actual lot size into account.

Now that we know the chances of the sample containing 0 defectives, all we have to do is repeat the process for 1 defective, then 2, then 3, then 4 and add the individual answers together. That's all!

I will lay out the chart for finding the probability that the sample contains 1 defective item. The rest I will leave as an exercise for the student! But, I will give you the final total so that you can check your work.

Sample Item Number	1	2	3	4	5	6	. . .	215	216	217	218
Item Condition	B	G	G	G	G	G	. . .	G	G	G	G
Number Items Left in Lot of Given Type	50	950	949	948	947	946	. . .	737	736	735	734
Number Items Left in Lot	1000	999	998	997	996	995	. . .	786	785	784	783
Prob	.0500	.9510	.9509	.9509	.9508	.9508		.9377	.9376	.9375	.9374

Now multiply all those 218 probabilities out and store the result on a piece of paper. Then move the Bad item from the first position to the second position and make the calculation again. And so on. After the Bad item in 218th position calculation is done, add all the position probabilities together and you have the chance that a sample of size 218 will contain 1 bad item. Now this sounds pretty time consuming. Indeed it would be if you didn't notice after the second or third calculation that all of the numbers you are getting are identical. That is, the probability for the sequence BGGGGGGGGGGGGG etc. GG is exactly the same as it is for the sequence GBGGGGGGGGGGGG etc. GG. And all you have to know is how many of the sequences there are - 218 in this case. There are 218 spots that the Bad item can fill - all of which have the same probability. Hence we only have to calculate the probability once and then multiply by 218. The probability for one sequence is about .00000022 and when multiplied by 218 gives the probability of .000048 that a sample of size 218 from a lot of size 1000 which contains 5% defective items will contain exactly 1 defective item.

Now if we were to add the .000003 (chance of 0 defectives) to the .000048 (chance of 1 defective), we would have the chance of accepting the lot IF the acceptance number was 1 - but for our sampling plan c = 4. That means of course that you have much more work to do - note the use of the words "you" and "work".

The next hurdle is the probability of finding two defectives in the sample. A sequence which will fill the bill is BBGGGGGGGG etc. GG. And you only have to calculate that probability once, and then multiply that probability by the number of sequences. It's the number of sequences question that I will address next. I'll assume that you have the probability of a particular sequence down pat.

Consider the three letters A, B, C. They can be arranged in the following ways. ABC, ACB, BAC, BCA, CAB, CBA - 6 ways. The first letter can be any one of the three. The second letter can be any one of the two remaining. And the third letter must be the one remaining letter. Let's put this in a table.

First Letter	A		B		C	
Second letter	B	C	A	C	A	B
Third letter	C	B	C	A	B	A

You can see the 6 arrangements. If there had been four letters - A, B, C, D then there would have been 24 arrangements. In general if you have n different letters, they can be arranged in:

n*(n-1)*(n-2)*...*3*2*1 = n! different ways

The symbol n! is read as "n factorial". Don't despair we're almost there.

If the set of n letters had n1 letters of one kind and n2 letters of a second kind and n3 letters of a third kind etc. up to nk letters of the kth kind, then the number of arrangements would be fewer and can be calculated as:

Number of arrangements =
n! / (n1! * n2! * n3! etc. * nk!).

For our 2 Bad, 216 Good situation the answer would be:

218! / (2! * 216!) = 218 * 217 / 2 * 1 = 23653.

Now aren't you glad you didn't write down all those arrangement probabilities?

What really happens in practice is that one uses calculators or computers to do this type of work. The "frequency Curve Comparison" module of Custom/QC allows you to get the answer without a whole lot of work. If you want to use the exact probabilities then you look up the answer in the "Hypergeometric Table."

The input looks like this:

Lot size	1000
Number defective in lot	50
Sample size	218
Number defective in sample	4

The output is: Probability of 4 or fewer defectives is .007597

Now this answer can also be approximated using the "Binomial Table".

The input looks like this:

Sample size	218
Fraction defective	.05
Number defective in sample	4

The output is: Probability of 4 or fewer defectives is .014323

In the example being studied, the sample size (218) is greater than 10% of the lot size (1000) and hence the approximation is not too good. But if you didn't have a computer to do the work and you wanted an answer, you probably would be satisfied to be this close. In fact, an approximation to the Binomial, known as the "Poisson Distribution" is often used.

The input for the Poisson is:

Mean (Sample size * % defective)	10.8
Number defective in sample	4

The output is: Probability of 4 or fewer defectives is .016517

You can see that the Binomial answer and the Poisson answer are quite similar. But because of the fact that the sample size is so large relative to the lot size, they both miss the Hypergeometric answer.

I don't know whether or not you still remember the question on the floor. The question is: How DO you get the other points on the Operating Characteristic (OC) curve for the plan? The answer to that question is: Use the computer and the Hypergeometric Table. There are some situations where the Binomial is the right table to use but I'll leave that to your friendly statistician.

The OC curve for the sampling plan we have been using as an example is shown below in Figure 20.

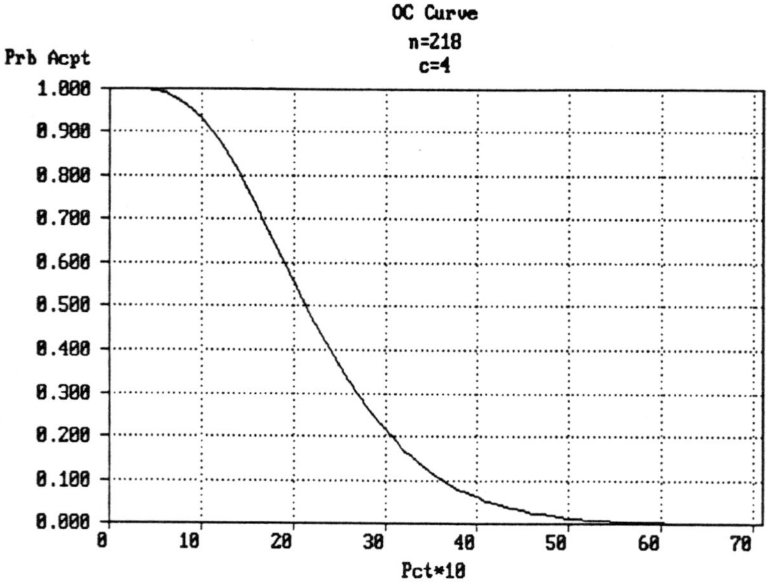

Figure 20

It probably seems to you that I spent a lot of time on Operating Characteristic (OC) Curves. There's a good reason for that: the OC curve is a complete summary of the risks involved in using the acceptance sampling plan of your (or your customer's) choice. There is one more reason:

!! EVERY SAMPLING DEVICE, including SPC charts, HAS AN OPERATING CHARACTERISTIC CURVE WHICH DESCRIBES THE SAMPLING RISKS ASSOCIATED WITH THE DEVICE. !!

So, you should at least be familiar with the concept.

The sampling plans we have been studying are generally known as "AQL" plans although I built the plan using both the Acceptable Quality Level (AQL) and the Rejectable Quality Level (RQL). The Mil Std plans are said to be "indexed by the Acceptable Quality Level (AQL)".

Another way of indexing plans would be by the Rejectable Quality Level (RQL).

Still another way to do the indexing is by a sampling plan performance measure known as "The Average Outgoing Quality Limit". These plans are called "AOQL" plans. Let's explore the AOQL concept.

AOQL–Average Outgoing Quality Limit

The Average Outgoing Quality Limit (AOQL) plans make the assumption that any lots which the plans reject will be 100% inspected and the bad items replaced with good items. The lots which a customer gets from an inspection scheme of this sort are, on the average, better than those which just go through the accept - reject schemes.

With this plan in effect, the customer gets two kinds of lots:

. Those which were rejected and hence are now 0% defective.

. Those which were accepted and hence have the

same quality level as before they were
inspected - I'm not going to pick nits about
the improvement which comes from replacing
the bad items found in the sample.

The Average Outgoing Quality (AOQ) for a plan is obtained from the Operating Characteristic (OC) curve as follows. The OC curve gives the Probability that a lot with a given % defective will be accepted. This value, which is on the y axis of the curve, is designated Pa - for Probability of Acceptance.

Average Outgoing Quality (AOQ) =
Probability of Acceptance (Pa) times the
lot % defective (p').
or, AOQ = Pa * p'

If a lot is 0% defective, then any sampling plan will accept it and the customer gets a lot which is 0% defective. If the lot is 100% defective, any sampling plan will catch it and again the customer gets a lot which is 0% defective - though where the good ones come from to replace the bad remains a deep secret in this case. So if you were to make a plot of the Average Outgoing Quality (AOQ) for your plan, it would start at 0% AOQ when p' = 0%, would rise up as p' increased, and then would return to 0% when p' = 100%. Thus, the AOQ curve reaches a peak somewhere between p' = 0 and p' = 100. This peak is known as the "Average Outgoing Quality Limit" or "AOQL" for the plan.

A plot of the AOQ curve for the sampling plan we have been using as an example is shown below in Figure 21.

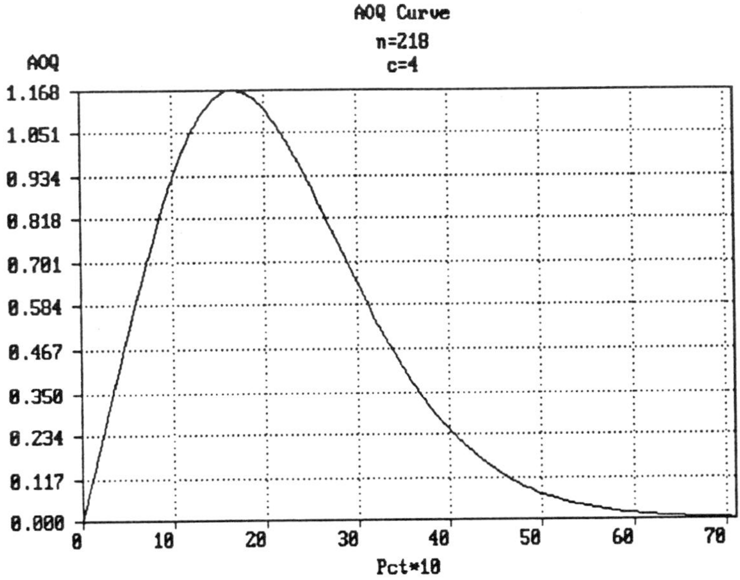

Figure 21

There's another topic in Acceptance Sampling with which you should really be familiar. The concept is that if you were to make a study of the quality of the lots which are actually produced in your plant, you would find that many of the lots you produce are 0% defective - and you have no worries. On the other hand some, few I hope, of your lots are 100% defective and any sampling plan will uncover this situation - and again you have no worries. It's the in-between quality levels in the lots that cause the trouble - they are sometimes accepted and sometimes rejected. There is much empirical evidence that the Lot Quality Distribution (LQD) has in general a sort of exponential shape. (See Taylor (10))

Were you to undertake the task of determining the Lot Quality Distribution (LQD) for your plant, this information could be used in conjunction with your sampling plan to describe the overall quality level that your customer will be receiving. For more information on this topic, look for articles having to do with Bayesian Sampling or Discovery Sampling. The net effect of the analysis will be to show you that, as you might expect, the quality level going to the customer is much better than the Average Outgoing Quality Limit (AOQL) might indicate. We have done some research in this area and would be glad to discuss the matter with you.

Now I have really gotten into sampling deeper than I thought I would. I hope you are still with me and ready to go into Volume II when it is completed.

Thanks for reading this material. If you have any suggestions for changes please let me know.

Appendix A

Formulas and Factors for Control Charts

FORMULAS:

Chart Type	Centerline	Lower Control Limit	Upper Control Limit
Averages			
Xbar & sigma	Xbarbar	Xbarbar - A1*Sigmabar	Xbarbar + A1*Sigmabar
Xbar & range	Xbarbar	Xbarbar - A2*Rbar	Xbarbar + A2*Rbar
Xbar & s	Xbarbar	Xbarbar - A3*Sbar	Xbarbar + A3*Sbar
Standard Deviations			
sigma	Sigmabar	B3*Sigmabar	B4*Sigmabar
s	Sbar	B3*Sbar	B4*Sbar
Ranges			
R	Rbar	D3*Rbar	D4*Rbar
Fraction Defective			
P	Pbar	Pbar - 3*Sigma P	Pbar + 3*Sigma P
		(Where Sigma P = Sqr(Pbar*(1-Pbar)/n))	
Defects			
C	Cbar	Cbar - 3*Sqr(Cbar)	Cbar + 3*Sqr(Cbar)

FACTORS:

	! Averages		!	Sigmas !		Ranges !		Estimate Sigma' !		
Subsize	A1	A2	A3	B3	B4	D3	D4	c2	c4	d2
2	3.76	1.88	2.66	0	3.27	0	3.27	0.56	0.80	1.13
3	2.39	1.02	1.95	0	2.57	0	2.57	0.72	0.89	1.69
4	1.88	0.73	1.63	0	2.27	0	2.28	0.80	0.92	2.06
5	1.60	0.58	1.43	0	2.09	0	2.11	0.84	0.94	2.33
6	1.41	0.48	1.29	0.03	1.97	0	2.00	0.87	0.95	2.53
7	1.28	0.42	1.18	0.12	1.88	0.08	1.92	0.89	0.96	2.70
8	1.17	0.37	1.10	0.18	1.81	0.14	1.86	0.90	0.96	2.85
9	1.09	0.34	1.03	0.24	1.76	0.18	1.82	0.91	0.97	2.97
10	1.03	0.31	0.98	0.28	1.72	0.22	1.78	0.92	0.97	3.08

Estimates of Sigma': Sigmabar/c2 Sbar/c4 Rbar/d2

Factors are taken from ASTM's STP 15D and from Grant and Leavenworth (5).

Appendix B

Average Run Lengths for Parabolic Mask on ComSum

Shift (sigma)	1.0	1.5	2.0	2.5	3.0
0.5	4.0	8.8	15.4	24.2	34.4
1.0	2.1	3.2	4.8	7.0	9.7
1.5	1.4	1.9	2.6	3.5	4.7
2.0	1.2	1.4	1.8	2.3	2.9
2.5	1.1	1.2	1.4	1.7	2.1
3.0	1.0	1.1	1.2	1.3	1.6

References

1. Dixon, W.J. and Massey, F.J., Introduction to Statistical Analysis, 4th ed. McGraw-Hill, 1983.

2. Shewhart, W., Economic Control of Manufacturing Products. Van Nostrand Company, Inc., 1931.

3. Aroian, L. and Levene, H., "Effectiveness of Quality Control Charts," Journal of American Statistics Association, December 1950, pp. 520-529.

4. Simon, General Leslie, "The Industrial Lot and Its Sampling Implications," Journal of Franklin Institute, vol. 237 (May 1944), pp. 359-370.

5. Duncan, A.J., Quality Control and Industrial Statistics, Revised Edition. John Irwin, 1959. Please see also the latest edition, 1986.

6. Duncan, A.J., "A Chi-Square Chart for Controlling a Set of Percentages," Industrial Quality Control, November 1950.

7. Gioia, L., "The Chi-Square Chart vs. X-Bar and R Charts," Unpublished, April 1986.

8. Hahn, G. and Cockrum, M.B., "Adopting Control Charts to Meet Practical Needs: A Process with Legitimate Batch to Batch Variability," General Electric Corporate Research & Development Report #84CRD314, December 1984.

9. Grant, E.L. and Leavenworth, R.S., Statistical Quality Control. McGraw-Hill, 1980.

10. Taylor, I.F., "Discovery Sampling by Attributes," North American Aviation, Inc., Columbus, OH., January 1956.

Index